William Pryor Letchworth

The Insane in Foreign Countries

William Pryor Letchworth

The Insane in Foreign Countries

ISBN/EAN: 9783337366322

Printed in Europe, USA, Canada, Australia, Japan

Cover: Foto ©berggeist007 / pixelio.de

More available books at **www.hansebooks.com**

THE INSANE

IN

FOREIGN COUNTRIES

BY

WILLIAM P. LETCHWORTH
PRESIDENT OF THE NEW YORK STATE BOARD OF CHARITIES

NEW YORK & LONDON
G. P. PUTNAM'S SONS
The Knickerbocker Press
1889

COPYRIGHT BY
WILLIAM P. LETCHWORTH
1889

TO

HIS EXCELLENCY

DAVID B. HILL

GOVERNOR OF THE STATE OF NEW YORK

AND TO

MY ASSOCIATE COMMISSIONERS OF

THE NEW YORK STATE BOARD OF CHARITIES

THIS

VOLUME IS RESPECTFULLY DEDICATED

CONTENTS

CHAPTER	PAGE
I. Introductory and Retrospective	1
II. England	15
III. Scotland	109
IV. Ireland	172
V. Continental Countries	194
VI. The Colony of Gheel	239
VII. The Provincial Insane Asylum of Alt-Scherbitz	279
VIII. Résumé	293
Index	365

LIST OF ILLUSTRATIONS

	PAGE
PINEL À LA SALPÊTRIÈRE	*Frontispiece*
CIRCULATING SWING .	9
BANDS AND CHAINS	10
NORRIS IN CHAINS .	19
DAY-ROOM, BROOKWOOD . .	76
COTTAGE HOSPITAL, BROOKWOOD	78
DINING-HALL, PRESTWICH .	94
INFIRMARY WARD, PRESTWICH .	96
CONVALESCENT WARD, PRESTWICH	98
DAY-ROOM, PRESTWICH . . .	100
BURNING AN INSANE WOMAN AS A WITCH .	110
TAKING LUNATICS TO DUBLIN	173
BELGIAN CAGE	197
PLAN OF THE COLONY OF FITZ-JAMES .	233
THE CHURCH OF ST. DYMPHNA . . .	240
PLAN OF THE PROVINCIAL INSANE ASYLUM, ALT-SCHERBITZ	280
RECEPTION STATION, ALT-SCHERBITZ	282
OBSERVATION STATION, ALT-SCHERBITZ	284
DETENTION HOUSE, ALT-SCHERBITZ	286
VILLA FOR PATIENTS, ALT-SCHERBITZ .	288
VILLA FOR THIRD-CLASS PATIENTS, ALT-SCHERBITZ .	290

PREFACE.

THE following pages are the outcome of an investigation of foreign charitable institutions, pursued without interruption through seven months, during which time special attention was given to the various kinds of provision made for the insane poor. The information thus collected is now published, in the hope of furthering the elucidation of a subject which materially concerns not only the State as a whole, but each one of its citizens.

For the purpose of securing fulness as well as accuracy, stenographic notes of visitations and interviews were made. By this means valuable opinions, expressed by distinguished specialists with whom the treatment of the mentally diseased has been a life-long study, have been carefully recorded. Avoiding, as far as possible, medical theories and controversies, the writer has striven to convey the results of his observations in plain, untechnical language. His aim throughout this whole inquiry has been to ascertain, from a practical point of view, what are the most advanced, the most humane, and the most economical methods of caring for the insane.

I feel it a pleasant duty to gratefully acknowledge the courtesies invariably extended to me by officials of different ranks in the countries visited. In many institutions considerable time was spent in making examinations, and the numerous details considered made the task of imparting information tedious; but a generous spirit was everywhere shown to meet the object of my visit. To the permanent

Secretary of the English Local Government Board, to the Commissioners and Secretaries of the English and Scotch Boards of Lunacy, and to the Superintendents of the various asylums inspected, I am under special obligations.

The publication of this work, most of the material for which was collected a few years since, has been delayed in consequence of a pressure of public duties. Meanwhile, however, correspondence has been maintained with superintendents of many of the institutions visited, and information kindly furnished by them, and the latest that could be obtained from other official sources are herein embodied.

W. P. L.

GLEN IRIS,
   Portageville P. O., N. Y.,
      1888.

INSANE IN FOREIGN COUNTRIES.

CHAPTER I.

INTRODUCTORY AND RETROSPECTIVE.

IN this age of high-pressure living, there is perhaps no subject of more general or more urgent interest than that of insanity in its relations to the State. No reflecting mind can be indifferent to the question of making proper public provision for the treatment and care of those afflicted with an insidious disease from which no measure of intellectual or physical strength or worldly prosperity affords any certain immunity—a disease, which, prone to feed upon excitement, finally transforms the noblest faculties of our race into a wreck so appalling, that in its contemplation the human intelligence becomes bewildered and dismayed. At no time in the history of civilization has the importance of this subject been more fully acknowledged; and probably at no time have influences contributory to mental derangement been more powerful than they are to-day. In America, where, from a variety of causes, there is so much mental activity, and where a condition of great prosperity attracts the surplus population of Europe, there are obvious reasons why the various problems relating to insanity should receive most careful attention. But before proceeding to the consideration of individual systems, it may be well to glance briefly at the past history of the treatment of the insane in different countries.

At the outset, we are reminded of the paradoxical saying of a distinguished writer, that the "ultimate tendency of civilization is toward barbarism." This, strange as it may appear, seems to have been exemplified by the degeneracy in the methods of caring for and treating the insane in different countries, from the days of antiquity down to the beginning of the present century, and even later. Among the ancient Egyptians, the priests, along with their supposed all-potent spiritual agencies, employed for restorative purposes such powerful aids as the influence of music and the beautiful in nature and in art, together with healthy recreation and agreeable occupation. Later, a member of a Greek medical school publicly condemned the excessive use of bodily restraint in the treatment of the insane. He advocated the importance of music and kindly treatment, as well as employment, and advised that patients be stimulated to self-regulation of their mental powers.

Centuries after those Eastern philosophers had passed away, during the Middle Ages, and down even to recent times, we find the idea prevalent in Europe of regarding the insane as possessed by demons, which must be cast out to effect restoration. This absurd superstition led to the adoption of cruel forms of punishment and even torture. The unhappy victims, objects of general abhorrence, were commonly cast into dungeons, where they were shamefully neglected. It seems not a little remarkable that the wonderful enlightenment of ancient times respecting this unfortunate class, should have been followed by such ignorance as is known to have prevailed during many succeeding centuries.

That great efficacy in the cure of mental diseases was

popularly ascribed to spiritual agencies, is abundantly shown, not only in legendary lore, but also in reliable historical records. Long pilgrimages were made to the shrines of particular saints, who were believed to have great influence in the work of mental restoration, and where, through exorcism and prayer, miraculous cures were claimed to have been effected. This practice was not confined to out-of-the-way places like Gheel, where St. Dymphna was held in holy reverence, nor was it restricted entirely to communities on the continent of Europe. The well of St. Winifred, St. Nun's Pool, and other wells in the British Isles, were resorted to by devout visitors, who were deeply impressed with a belief in the curative influence of these waters. At St. Nun's Pool it was the custom to plunge patients backward into the water and drag them to and fro until their excitement was subdued. Then they were taken to the neighboring church, where, if there were signs of recovery, thanks were offered for their deliverance; if otherwise, the hydropathic treatment was continued while there remained any hope.

In Ireland, there was Glen-na-galt, or the "Valley of the Lunatics," beautifully situated in County Kerry, not far from Tralee. To this vale it was believed that every lunatic would eventually gravitate if left to himself. The process of cure consisted in drinking the cooling waters and eating the cresses that grew beside the spring.

In Scotland, many wells were traditionally celebrated for similar wonderful properties. It is recorded, that, not many years since, lunatics were denuded and thrice dipped at midnight in Lochmanur, in the far north of Scotland. Moreover, in the western highlands, the poverty-stricken inhabitants, shut out, as it were, from the influence of town or city, are still, to some extent, swayed by superstitious notions that have been handed down from father to son.

Whittier alludes to the popular belief in the remarkable virtues of the well of St. Maree. After referring to the soothing and restorative effects from bathing the fevered brow therein, he goes on to say:

> "That holy well of Loch Maree
> Is more than idle fable."

Farther south, in a picturesque district of the midlands of Scotland, we learn, on the authority of Sir Walter Scott, that several wells and springs were religiously dedicated to St. Fillan, and that down to very recent times they were places of pilgrimage and offerings, because the waters were believed to be a potent means of cure in cases of madness.

Far into the Middle Ages, the practice of medicine was left to the monks, who ministered alike to soul and body. In the days of the ancient Greeks, psychology was recognized as a department of medical study; but the ignorance of the Dark Ages affected medical, as, indeed, every other progressive science, and before the general dissemination of education had time to show its effects, many spurious remedies and crude superstitions were incorporated into the healing art. We may smile at the lack of knowledge which induced our forefathers to administer, as remedies, "open drinks," in which bitter herbs, ale, and holy water were essential ingredients; at the clove-wort wreath enjoined to be worn "when the moon was on the wane"; and at the prayers addressed to the periwinkle at different stages of the moon. The following was a more drastic prescription: "In case a man be lunatic, take a skin of mere-swine (in other words a sea-pig or porpoise), work it into a whip and swinge the man therewith; soon he will be well." Some of the old-fashioned remedies are, however, part of the pharmacopœia of the present day. Hyoscyamus is still administered in disturbed cases.

Throughout the sixteenth century, and far advanced into the seventeenth century, demonology and witchcraft were generally believed in, and he was a bold man who ventured to contradict the orthodox opinions of the period. These superstitious beliefs were not confined to the illiterate. No less a personage than James VI. of Scotland, afterwards king of England, wrote a learned dissertation on " Demonology," in which he displayed, perhaps, more zeal than discretion, by contending that evidence not admissible against other offenders should be accepted against so-called witches; that marks about their persons should be diligently searched for and " pricked with a long needle," to ascertain whether they were insensible to pain (such insensibility being unmistakable evidence of the satanical mark, or " seal of the devil"); and that their bodies should be floated on water, as further aids in discovering the guilty. Toward the close of the sixteenth century, certain medical authorities began openly to rebel against these accepted doctrines, whereupon they were severely censured by the royal James for what he was pleased to term their " damnable opinions."

As recent as 1716, a woman and her daughter were sentenced to death at Huntingdon, by an English bench of judges, for "selling their souls to the devil." Many such poor people, it is hardly necessary to say, were simply mentally deranged, and had they been so fortunate as to live in recent times, they would have been humanely cared for in some lunatic asylum. Such deplorable ignorance and superstition, however, were by no means limited to England. In Scotland, also, the practice was common of putting the insane to death, under the belief that they were witches and possessed of an evil spirit. Instances of similar superstitions might be multiplied from the records of our own and other countries.

The first asylum of which we find mention in history is one said to have been erected by the monks at Jerusalem in the latter part of the fifth century. The learned monks appear to have been the principal pioneers of this charity during the Middle Ages, though there are evidences that provision for the insane was not entirely neglected by the Mohammedans. Among European countries, in the fifteenth century, Spain, then a centre of learning, seems to have taken the lead in providing for this class. The treatment adopted in Spain, as elsewhere, was, however, cruel in the extreme, and was based on the general belief that the insane were possessed by evil spirits. Many were burned to death, others were scourged and tortured in the vain hope of expelling the demons and liberating the victims from the power of the Prince of Darkness. It should be acknowledged, that, amidst all these prevailing errors, the monks at Saragossa had the first faint conception of rational open-air treatment for the mentally deranged to be found in modern times.

So great a revolution has taken place within the past few generations in the methods of caring for the insane, that we are scarcely able to realize that a hundred years have not elapsed since persons of unsound mind were treated worse than wild beasts,—everywhere kept under bolt and bar, or heavily manacled in cells and dungeons, the poisoned atmosphere of which not only prevented cure, but hastened death. But a little earlier, in some cases they were exhibited in cages to the public, at fixed rates, and were irritated and tormented to gratify a morbid and vulgar curiosity. As a result of this treatment, many who originally were not beyond hope of recovery, became permanently deranged. Those who were not deemed dangerous to the public safety were left to roam about the country in a neglected and pitiable condition. It is difficult to conceive that a century has not passed since the Herculean labors of Pinel were needed to loosen the chains of the miserable occupants of the Bicêtre in Paris—an act which, as is well known, revealed to the world the striking superiority of kindly treatment over the torture, worse than death, to which insane men and women were at that time subjected in different parts of Europe. This marks so important an era in the history of the treatment of the insane, that no excuse is necessary for adverting to some of the attendant circumstances of that great achievement.

Pinel, who had attained some distinction as an alienist, was appointed, in 1792, to fill the post of superintendent of the Bicêtre, which then contained upwards of two hundred male patients, believed not only to be incurable, but entirely uncontrollable. The previous experience of the physician, here stood him in good stead. He had been a diligent student of the authorities of his own and foreign countries on diseases of the mind, and in his earlier years had been appointed by the French government to report on the condition of the asylums at Paris and Charenton. On assuming the oversight of the Bicêtre, he found fifty-three men languishing in chains, some of whom had been bound for a great number of years. These were regarded by the authorities as dangerous and even desperate characters; but the sight of men grown gray and decrepit as the result of prolonged torture, made a very different impression on the mind of Pinel. He addressed appeal after appeal to the Commune, craving power to release, without delay, the unhappy beings under his charge. The authorities tardily and unwillingly yielded to the importunity of the physician. An official, who was deputed by the Commune to accompany the superintendent and watch his experiment, no sooner caught sight of the chained maniacs than he excitedly exclaimed: "*Ah, ça! citoyen, es-tu fou toi-même de vouloir déchaîner de pareils animaux?*"[1] The physician was not to be deterred, however, from carrying out his benevolent project, and did not rest satisfied until all of the fifty-three men had been gradually liberated from their chains. Singular as it may appear, the man who had been regarded as the most dangerous, and who had survived forty years of this severe treatment, was afterwards known as the faithful and devoted servant of Pinel. The reforms of Pinel were not confined to the Bicêtre, an establishment exclusively for men, but extended to the Salpêtrière, an institution for women. There is, perhaps, no more touching event in history than that of this kind-hearted and wise physician removing the bands and chains from the ill-fated inmates of this place of horrors.

The monstrous fallacy of cruel treatment once fully exposed, the insane came to be looked upon as unfortunate human beings, stricken with a terrible disease, and, like other sick persons, requiring every aid which science and benevolent sympathy could provide with a view to cure. Governmental inquiries were instituted with a view to the attainment of better treatment, and in different countries, almost simultaneously, the provision of suitable and adequate accommodation for the insane was declared to be a State necessity.

After scientific thought had at length been turned to the cure of insanity, many new and strange devices for its treatment were adopted, which underwent, one after another, the test of experiment. Some of these, in our day, are considered not only absurd, but highly injurious, although

[1] Ah, now! citizen, art thou mad thyself to desire to unchain such animals?

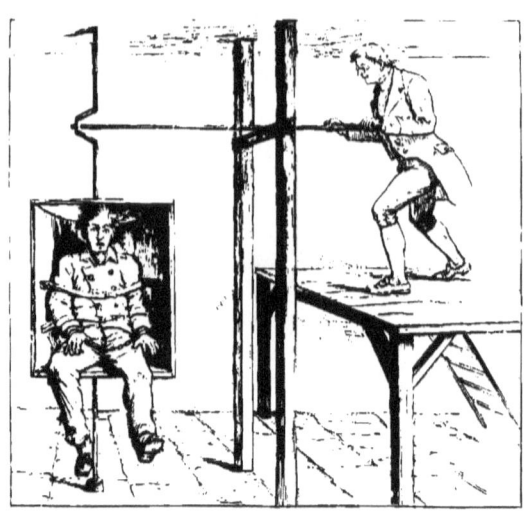

THE DOCTOR THINKS THAT "NO WELL-REGULATED INSTITUTION SHOULD BE UNPROVIDED WITH THE CIRCULATING SWING," 1818.

originating with men distinguished for their learning, who regarded them at the time as most reasonable and efficacious. An illustration of one of these devices is found in an invention of the medical superintendent of a foreign insane asylum. It is described in a work published by him in 1818, in which he gives his observations and experience as follows:

"The circulating swing erected in our asylum appears to be an improvement on the model suggested by Doctor ———. It is worked by a windlass, and capable of being revolved a hundred times in a minute; but can with ease be regulated to the degree best suited to the intent. It is now adapted for one person only instead of four, as had been at first contrived; the same movement being seldom admissible for more than one patient. To the body of the machine is affixed the apparatus for the horizontal position, which, when necessary, may at a moment be accommodated to the purpose.

"Powerful as this contrivance has hitherto proved, still, in some cases, where its influence was much sought for, it has had but trivial effect, though put in motion to its full extent. The idiots of the establishment have been permitted to use it for amusement without any inconvenience, and the strictly insane also, during their intervals, with equal satisfaction. The latter, however, on the return of the paroxysms, were found incapable of resisting its most gentle rotations for five minutes in continuance.

"In several cases of continued insanity of long standing, where the swing had been employed as a last resource, I have been most agreeably surprised at the unexpected alteration which was effected after a few trials. In some, who, from their disposition to violence, and who from necessity were closely confined to solitary apartments, it had so far succeeded as not only to render them easy of access, but also to induce kind and gentle manners, effecting in the end the most willing service in the daily occupations of cleansing, and attendance on the sick. It may be proper to remark that those persons previously to any such amendment were invariably affected from the disturbance occasioned by the swing, with a smart fever of eight or ten days duration, and from which the favorable occurrence here alluded to, seemed to have arisen. I cannot undertake to say that, where the disease had assumed the chronic and uninterrupted form, any one instance of complete recovery has as yet followed from its use; yet as it has thus far established its utility, it is to be presumed that no well-regulated institution intended for the relief of the insane will be unprovided with a swing of a proper construction as a curative expedient eminently adapted to the purposes for which it is so particularly recommended."

The same learned and worthy doctor describes a belt and a leg chain of his invention, which he considered great remedial agencies, as by the use of these appliances, patients under treatment, however violent, could take daily exercise.

"To obviate the supposed inexpediency of freeing maniacal patients from close confinement, and to provide against the possibility of their using unrestrained violence during this indulgence, a broad strap, made of strong leather, has been provided, which embraces the body directly under the axillæ. At the sides of this strap are affixed other narrower straps, fitted to hold each arm, so as to be buckled behind, and by which means they are effectually secured. The main strap is closed behind with loop-holes and rings, through which a small iron pin passes, stayed at top, and fitted at bottom for a padlock. To prevent the body strap from slipping down, a loose circular band of leather extends from the front edge over the head and lodges broadly on the shoulders, which is also secured behind by a loop-hole and small strap attached to one of the rings through which the pin is directed. By this method the whole is preserved from being shifted, and it most completely prevents the use of either hand for any injurious purpose. The object of exercise is, however, sufficiently attained, for which the open galleries of the institution are very well adapted, till at length the patients are permitted, under the same restrictions, to pass into the open air.

"A determined perseverance in this plan, sometimes under very remarkable circumstances, has had its full reward. Gangrene of the lower extremities is no longer known in the asylum. No patient is allowed to remain confined to bed during the absence of fever, or to his apartment, unless on the principle of occasional punishment."

SHOWING HOW BANDS AND CHAINS IN THE PAST WERE MISTAKENLY USED AS AUXILIARIES TO CURE.

Perhaps some of the means employed in the early history of curative treatment do not seem more ridiculous to us than some of the remedies that we employ, will, in after time, appear to our successors. It would, therefore, seem the part of wisdom to be not over-confident in our present methods of treatment, but to look, with careful inquiry, to new sources of relief that promise good results.

Pursuing the history of the treatment of the insane and turning our attention from the Old World to the New, we find that, in America, many distinguished philanthropists have, during the present century, labored with zeal and energy to ameliorate the condition of this class; and in the alleviation of suffering humanity they have performed a noble work.

The initial point in hospital treatment of the insane, in this country, was in 1750, when there was projected in the " Province of Pennsylvania " a small hospital, to be located at Philadelphia, in which the principle of treating the patients as " sick persons " was recognized. The Act incorporating it, passed in 1751, was entitled " An act to encourage the establishing of a hospital for the relief of the sick poor of this province, and for the reception and care of lunaticks." This small institution, begun purely as a curative establishment, in what had been a private dwelling, was, at its outset, dependent on private subscriptions. The services of the visiting physicians were gratuitous, and medicines were also furnished free of charge. At this time, punishment, in one form or another, was, in general estimation, regarded as only second in importance to bleeding, purging, and dosing. Even when, in 1783, the philanthropic Dr. Rush resolved to relinquish the whips and chains of his day, he still adhered to " mild and terrifying modes of punishment," and, on paying a visit to an insane man, deemed it one of the first requisites " to look him out of countenance."

The first State asylum erected in this country, and exclusively devoted to the insane, was opened at Williamsburg, Virginia, in 1773. The building was erected at a cost of £1,070, and supplied, also, the wants of Kentucky, which had not then taken rank as a State. The object of this institution, as defined by an act of the Legislature, was the "cure of those whose cases are not become quite desperate, and for restraining others who may be dangerous to society."

In 1817, the Society of Friends in America, stimulated by reports of the good work then being accomplished at the York Retreat in England, purchased fifty-two acres of land at Frankford, near Philadelphia, and there opened a hospital, the object of which is best expressed in the language of the projectors, "that the insane might see that they were regarded as *men* and *brethren*."

The organization of the McLean Asylum, at Somerville, Mass., in 1818, begun in a building that had been a private dwelling, was another important event in respect to recognizing the principle of State supervision; part of its board of trustees, charged with weekly visitation, being appointed directly by the Governor of the State. Then followed a period of great reforms, in which men of eminence in the medical profession, and philanthropists like Miss Dix, received hearty legislative coöperation in their work. In this way institutions for the insane in the United States have, since 1829, been increased more than tenfold.

In the State of New York, we find that the first legislative recognition of the necessity of asylum provision for the insane was in 1801, when an act was passed providing that lunatics might be admitted into a department of the New York Hospital, the committing authorities paying for their maintenance. State aid was granted in furtherance of this object. The records of the Hospital, however, show that insane patients had been admitted as early as 1797. In 1808, a separate building, known as the South Hospital, was also opened for this class. Aided still further by special grants from the Legislature, a branch of the institution devoted exclusively to the insane was opened in 1821, under the name of the Bloomingdale Asylum.

In the period which elapsed from 1830 to 1850, great and rapid advances were made throughout the United States in methods of caring for the insane. The reforms then accomplished attracted the attention of Europe, and it may be said, without any egotism, that they were in advance of contemporary progress in other countries. But excellence can be maintained only by continual progress. It has been freely asserted, that, from the middle of the century up to a comparatively recent period, this country failed to keep pace with the spirit of reform elsewhere manifested. It is gratifying to reflect, however, that during the past decade the advances made here have been as rapid as in the period between 1830 and 1850. Should this improvement continue, it seems not unreasonable to expect that the time is not far distant when the methods of caring for the insane in the United States will reach a higher standard than those of any other country.

Among the criticisms of British specialists who have visited asylums in this country are the following: That some, even of our most modern institutions, are unnecessarily prison-like in construction; that restrictions which are positively hurtful are still imposed upon patients; and that, in other respects, the asylum systems of the United States are more artificial and less natural or homelike than the systems of some other countries. It is not to be supposed that these criticisms apply to all of our asylums, the statements being entirely and admittedly inapplicable to some of our best-managed institutions. But, whatever may be the measure of application of these strictures, they have been advanced by men, who, in their own country, are regarded as authorities, and should receive careful and candid consideration.

The writer does not expect that the succeeding pages will set at rest controverted points that have been raised respecting the care and treatment of the insane ; but he hopes that some of the illustrations given of foreign methods will prove helpful to those concerned in the adjustment of questions now uppermost in the public mind and bearing on a subject of more than national importance, affecting, as it does, the whole human family.

CHAPTER II.

ENGLAND.

THE history of the treatment of the insane in England is the history of a protracted struggle between the forces of a humane and enlightened civilization on the one hand, and of a formidable and unyielding conservatism on the other. Before adverting to the high position which many of the English asylums have attained, it may be well to recall briefly the gradual steps by which this was reached. In 1547, a monastic institution, established some three hundred years earlier, and now known as Bethlem Hospital, was converted by Henry the Eighth into an asylum, or more correctly speaking, a "dungeon-house," for furious lunatics, in which capacity it became popularly known as Bedlam. This is the first public institution in England devoted to the insane of which we have any record. It was only large enough to accommodate some sixty patients, and in 1675 it was removed to Moorfields, where a building for one hundred and fifty-two inmates was erected. Extensive additions were made to the hospital from time to time, but in 1812 it became necessary to again change the site, and the more modern structure was built in St. George's Field.

The earliest recorded English legislation bearing directly upon the insane poor was in 1744, in George the Second's time, when any two justices were empowered to order the arrest of pauper lunatics found at large, and have them chained and locked up in " some secure place " within their parish of settlement. Up to this period, such provision as had been made for the insane was, however, mainly that of custodial care for those who were dangerous to be at large.

Many of the mentally deranged were to be found wandering about the country in a condition of pitiable neglect—a source of apprehension and annoyance to the public, and an unwelcome addition to the mendicancy of the time, which is thus graphically sketched by Shakespeare :

> " The country gives me proof and precedent
> Of Bedlam beggars, who, with roaring voices,
> Strike in their numb'd and mortified bare arms
> Pins, wooden pricks, nails, sprigs of rosemary ;
> And with this horrible object, from low farms,
> Poor pelting villages, sheep-cotes, and mills,
> Sometime with lunatic bans, sometime with prayers,
> Enforce their charity."

The first asylum in England in which the curative principle is distinctly discernible, is that of St. Luke's, erected by the subscriptions of a few charitable persons in London, in 1751. About the close of the eighteenth century, there were in England not more than five public asylums—mostly endowed—and only a few private institutions.

Down to the year 1770, Bethlem Hospital, which was then, as now, managed by a board of governors, derived a considerable revenue in fees from the public exhibition of the insane, heavily manacled and chained, or kept in cages of iron. It would appear that the visitations on the part of the curious were the only kind of inspection to which these institutions were subject ; and, until the year 1853, the governors of Bethlem Hospital resented, and successfully resisted, governmental supervision. Influenced probably by a national calamity, in that the reigning sovereign, George III., had several times been attacked by insanity, public attention was in his day fairly aroused to the necessity for humane and more perfect asylum accommodation and supervision. In 1774, an act was passed directing the Royal College of Physicians in London to appoint a commission, consisting of five of its Fellows, to visit and license the "mad-houses," as they were then termed, situated within the cities of London and Westminster, or within seven miles thereof. The Act further directed that similar duties should be performed by the magistrates in their respective districts.

The custody of the estates of idiots, formerly intrusted to the feudal superior, was, in the reign of Edward II., vested in the King, and proved no inconsiderable source of revenue to the crown. A distinction was made by law as between idiots, or "natural fools," and lunatics. Blackstone defines an idiot as "one who hath had no understanding from his nativity, and therefore is by law presumed never likely to attain any"; and a lunatic as "one who hath had understanding, but by disease, grief, or other accident hath lost the use of his reason." In the case of the former there was forfeiture of the surplus profits of the estate during the lifetime of the idiot, after supplying him with necessaries; in the case of a lunatic, the crown acted as trustee, protecting his person and property, and accounting to him, if he recovered, for all profits received, or to his representatives in the event of his death. In the case of idiots the property was preserved for the benefit of the heirs. Thus appears to have originated that care or guardianship of estates of idiots and lunatics which is still maintained by the Court of Chancery.

But whatever of humane care was manifested, it was apparently confined, as Chancery care is now confined, to those possessed of estates. Pauper and criminal lunatics, violent, and dangerous to be at large, were, without ceremony, relegated to the gloomiest confines of jails and workhouses.[1] Even in what were termed asylums, the accommodation included dark, damp cells, where, as if to accelerate death, iron cages, manacles, wristlets, and massive chains were unrelentingly employed, lacerating and discoloring the flesh of the unhappy victims. A grave abuse connected with these receptacles for the insane, lay in the fact that they were resorted to by the powerful and unscrupulous as conveniences for getting rid of any relative who might happen to stand in the way of their selfish aims. The poorer classes were crowded to suffocation in dungeons, bled profusely, lashed into a state of complete exhaustion, or, as too frequently happened, were brutally kicked to death by some unfeeling attendant, who, free from supervision, knew that he might with impunity inflict fatal injury upon any one of the patients committed to his charge. This fearful state of things is amply substantiated by competent authorities.

The late Dr. R. Gardiner Hill says:

"In the early part of the present century, lunatics were kept constantly chained to walls in dark cells, and had nothing to lie upon but straw. Some were chained in dungeons, and were gagged, outraged, and otherwise abused. The keepers visited them whip in hand, and lashed them into obedience; they were also half-drowned in baths of surprise, and in some cases semi-strangulation was resorted to. The bath of surprise was so constructed that patients, in passing over a trap-door, fell in. Some were chained in wells, and the water made to rise until it reached the patient's chin. . . . Patients in a state of nudity, women as well as men, were flogged at particular periods, chained, strapped and fastened to iron bars, and even confined in iron cages."

The cruel treatment of the times is truthfully illustrated by the oft-quoted case of Norris, who was an inmate of Bethlem Hospital before that place came to rank, as it now does,

[1] *Workhouse* in England is synonymous with *poorhouse*.

NORRIS IN CHAINS

among the better class of English asylums. It appears that Norris, in an outburst of passion to which the nature of his infirmity rendered him momentarily liable, had struck the apothecary of the asylum, and for this offence was kept in irons for twelve years! The refinement of torture was in this instance reached by keeping the victim in such a position that he could not stand upright! A reliable authority thus particularly describes the mode of his restraint:

"A stout iron ring was riveted round his neck, from which a short chain passed to a ring made to slide upwards or downwards on an upright massive iron bar inserted into the wall. Round his body a strong iron bar about two inches wide was riveted; on each side of the bar was a circular projection, which, being fastened to and inclosing each of the arms, pinioned them close to his sides. The waist-bar was secured by two similar bars, which, passing over his shoulders, were riveted to the waist-bar both before and behind. The iron ring round his neck was connected to his shoulders by a double link. From each of these bars another chain passed to the ring on the upright iron bar. His right leg was also chained."

That cruelty of the kind above described should have been possible as recent as the year 1815—twenty-three years after Pinel's great reform in France, and after Tuke's introduction of humane principles at the York Retreat—is almost beyond belief. Yet authentic records show that Bethlem Hospital was by no means the only one where the patients were brutally treated. It was the discovery of similar scandalous practices at the York Asylum—established by general subscription in 1777—that led William Tuke, in behalf of the Society of Friends, to found the famous Retreat at York, long the only English institution of its kind in which chains, leg-locks, and hand-cuffs were totally discarded. The circumstances leading to the establishment of this novelty in asylum management were as follows: The Society of

Friends in 1791 sent one of their number, a female patient, to the York Asylum, when it was learned, that, by the rules of the institution, none of her relatives would be permitted to see her. There was reason to fear that the reports then in circulation about the inner workings of this place of concealment and confinement were not without foundation, and as the result of investigations, the Society resolved to erect an independent hospital where moral treatment should have precedence over mechanical methods.

The Retreat was opened accordingly in the year 1796. One of the first patients admitted was a man who had been chained for twenty years. Under the new *régime* of kind and more intelligent treatment, he was at first restricted only by the occasional use of arm-straps, and he soon consented to wear his clothes, becoming in this and other respects decent and orderly in his habits. The success attending these efforts need not be dwelt upon here.

Meanwhile, the unenviable notoriety of the York Asylum increased rather than diminished. Regular official inspection there was none to confirm or refute the charges which were in circulation regarding the prevailing usages. There is hardly an atrocity that could be named, including sexual outrage, that is not alleged to have been perpetrated upon the helpless victims then incarcerated in the gloomy cells of that prison-like structure. Its stone floors were begrimed with filth, and its round air-holes, only eight inches in diameter, exhibited the merest mockery of ventilation. In consequence of complaints made by members of the Society of Friends, an official inquiry was instituted into the management of this and similar mis-named "public" asylums. It is authoritatively asserted of the York institution that scarcely had the inquiry terminated, when the wing of the asylum which contained the worst cases was purposely set on fire to destroy evidences of hideous cruelty, and several of the patients were burned to death. Statistics of the institution were found to have been falsified; 221 patients were advertised as dead, whereas the number ought to have been 365. Among other recorded instances of neglect, it is stated that the janitors took no measures to protect the insane under their charge from the attacks of rats, which allowed the patients neither rest by day nor sleep by night.

In 1807, a select committee of the House of Commons inquired into the condition of the lunatic poor, and in the following year an act was passed providing for the erection of properly equipped public lunatic asylums, to be built at the expense of the counties or boroughs throughout England and Wales. This and a few subsequent amending acts left it optional with the justices or magistrates to provide such asylums; but, in 1828, only twelve counties had made the requisite provision. By the year 1841, only six other counties had complied by furnishing the desired accommodation. Meantime, parliamentary investigations, of which there were several between 1808 and 1844, revealed a very unsatisfactory state of things. Many of the asylums had no medical attendant; many, both in and out of London, were seldom visited, and great abuses were common. In 1828, an act was passed empowering the Home Secretary to appoint, every year, fifteen commissioners—five being physicians—for the purpose of licensing and visiting the metropolitan district. Four years later the Lord Chancellor was authorized to appoint annually the Metropolitan Commissioners in Lunacy, and finally, in 1842, these commissioners were empowered to inspect all the public and private asylums throughout England and Wales. They made a searching inquiry into the condition of these institutions, and their report made to Parliament in 1844 proved of great value in shaping subsequent legislation.

The published accounts which had been periodically emanating from the York Retreat, were influential in confirming the doctrine laid down by Pinel, the great French reformer, that severe forms of restraint are hindrances, and not helps, in the treatment of the insane. Yet so strong was the prejudice against reform, that, when Dr. Hill attempted it in the Lincoln Asylum in 1838, he was forced to resign his office of superintendent. In the following year Dr. Conolly was appointed to the large metropolitan asylum of Hanwell, in Middlesex. Having obtained the reluctant consent of the authorities to what was believed to be a dangerous experiment, he boldly adopted those principles of non-restraint which had been already firmly established by Pinel and Tuke. So marked was his success in this large public asylum set apart for disturbed cases, that Gloucester and many other provincial public institutions, including even conservative Lincoln, became converts to the new method. The substitution of humanity for brutality gradually extended, though much work had yet to be done. It was commonly considered as late as 1842, that old jails required but little alteration to make them fit receptacles for the insane.

The nature of the provision made for the insane about this time and prior to 1844, was described by the Earl of Shaftesbury, Chairman of the English Board of Commissioners in Lunacy, before a parliamentary inquiry, in 1877, in these words:

"I recollect I used to see as many as thirty, forty, or fifty patients chained to the wall. I never knew an attendant go about who had not leg-locks and hand-locks to his waist, which were applied without remorse. I do not mean to say that people in those days were less humane than we are now, but they were ignorant, and many of them thought that a madman was a creature so devoid of sense and feeling, that he might be treated not only as a beast, but worse than a beast. I remember, as to the White

House Asylum in Bethnal Green, it came out in the evidence, and the physician who was at the head of it never denied the statement, and as a matter of course he did not think there was anything inhuman in the fact, that, on Saturday night, from 200 to 220 patients were chained down in their cribs and never visited again until Monday morning. There was a crust of bread and a cruse of water put beside them, and they were left in their filth. Pauper patients did not then sleep in beds as they do now. Go and see them now in their dormitories, you would have no objection to sleep there yourselves; but they were then put in cribs with straw—put in naked,—and there were holes in the bottom to let the urine pass; the rooms were all constructed with flags or bricks, sloping down in the centre like a filthy stable with a drain in the middle, and there they were lying for hours."

Among the crying defects of the time were the improper crowding together of patients, a want of medical care and proper nursing, unnecessary coercion and restraint, and an absolute dearth of occupation, recreation, or amusement. The methods of restraint in common use at this time included iron hand-cuffs connected with chains to a leather waist-belt; leathern hobbles locked around the ankles, which permitted the patient to shuffle his feet, but impeded his movements so that he could not walk; iron hand-cuffs with chains which passed through fastening locks at the sides of heavy "tub-bedsteads" filled with straw; boots made of ticking, with rings and chains which passed through fastening locks at the bottom of the bedstead.

By 1845 it had become evident that the law allowing magistrates, at their option, to make asylum provision was not beneficial in its practical operation. Accordingly, by an act passed in that year it was made incumbent upon the justices of every county and borough, within three years, to obtain or provide the necessary accommodation for their insane. They were allowed to make this provision either separately or jointly with other counties or boroughs. This act, modified by subsequent enactments to be hereafter referred to, forms the basis of the existing system. The new law superseded all previous legislation on the subject; appointed for life eleven Lunacy Commissioners, three of whom should be physicians and three barristers; and directed that all future vacancies should be filled by the Lord Chancellor. The six professional commissioners were to be the only paid members, and they were specially charged with the work of visiting. A Secretary to the Commission was also appointed, and all took the oath of fidelity and secrecy as administered by the Lord Chancellor.

The commissioners, acting under a common seal, were empowered to elect from their number a permanent unpaid chairman; to make rules for their guidance in carrying on their work; to visit and supervise all asylums certified to receive the insane throughout England and Wales. They were authorized to inspect the books and documents of the institutions visited to see that they were regularly kept; in short, to inquire into all matters and things pertaining to the physical or mental condition of the insane. Any keeper or superintendent withholding the required information was made liable to pains and penalties for misdemeanor. They were required to make an annual report to the Lord Chancellor, for presentation to Parliament.

In 1852, notwithstanding much previous legislation on the subject, several counties were still without asylums, and only four boroughs had provided separate accommodation. It was therefore found necessary to pass a still more stringent act in 1853, making it compulsory on justices of every county or borough to provide for the wants of pauper lunatics in their districts, either separately or by agreement with others. The word "lunatic" was expressly defined to " mean and include every person of unsound mind, and every person being an idiot." The justices were now required to appoint committees of their number to erect asylum buildings or provide needful accommodation ; and it was declared that counties and boroughs neglecting to so provide or contract for their pauper lunatics should be annexed (for the purposes of the Act) by the Home Secretary to the adjacent counties. To meet the case of crowded localities like the metropolis, where lands and accommodation might be difficult to procure within a given area, power was granted to authorities to erect and manage asylums outside their own county should circumstances so require. The cost of the new arrangement was directed to be defrayed by a county tax. Borrowing powers were also conferred upon local authorities to meet necessary expenditures.

The primary intent of the statute in creating county and borough asylums was to provide for pauper lunatics. Other lunatics, or the pauper lunatics of other counties or boroughs, may not be admitted to any asylum of this kind until all the pauper lunatics of the county in which it is situated, or the counties or boroughs contributing to it, if any, are accommodated.

The management of the county and borough asylums is entrusted to a " Committee of Visitors " appointed annually by the justices of the counties and boroughs. This committee, of whom three constitute a quorum, are empowered to appoint a paid clerk who may also be the clerk of the asylum, and in whose name they may sue and be sued. The committee are likewise empowered to appoint a resident superintendent, chaplain, treasurer, and such other officers and servants as they may think fit. They have also power to discharge the same. Should a nurse or attendant be discharged, the cause for so doing must be reported to the Commissioners in Lunacy. Subject to the approval of the Home Secretary, the committee may frame rules and regulations for the government and management of the asylum under their charge, and not less frequently than once during every two months as many as two members of the committee are required to visit the asylum together and make a thorough examination of every part, and to see, "if circumstances will permit," every patient therein. They are to examine all commitment papers received since their last visitation, also the books of the asylum, and they are required to enter in a book kept for the purpose at the asylum their impressions of the condition of the asylum, the patients, etc. The Committee of Visitors report annually to the justices of the county or borough, and a copy of this report must be sent by them to the Commissioners in Lunacy.

The statute respecting the maltreatment of asylum patients provides that—

"If any superintendent, officer, nurse, attendant, servant, or other person employed in any asylum strike, wound, ill treat, or willfully neglect any lunatic confined therein, he shall be guilty of misdemeanor, and shall be subject to indictment for every such offense, or to forfeit for every such offense, on a summary conviction thereof before two justices, any sum not exceeding twenty pounds nor less than two pounds."

It is required by the statute that every house in which more than one patient is kept, whether private or pauper, must be licensed. The district lying within the distance of seven miles from any part of the cities of London and Westminster, or the borough of Southwark is designated as the "immediate jurisdiction" of the Commissioners in Lunacy, and within this district licenses are granted by them. Elsewhere such licenses are issued by the justices of the county or borough within which the house is situated.

The justices are also required to appoint three or more justices and one physician or more to act as local visitors to the licensed houses within their several localities. The names of these visitors must be forwarded to the Lunacy Commissioners. All houses licensed for one hundred patients or more must have a resident medical officer. If, in any house licensed for less than one hundred patients, there be no resident medical officer, such house must be regularly visited by a physician, the frequency of his visits being regulated according to the number of patients for which the house is licensed.

It is also required that any hospital or part of a hospital or other house or institution (not being an asylum) wherein lunatics are received and supported wholly or in part by voluntary contributions, or by any charitable bequest or gift, or by applying the excess of payments of some patients for or towards the support or benefit of other patients, must be registered before patients may be received therein. Should the superintendent of such an institution fail to make application for registry to the commissioners, he is liable to a fine of £20. After registry, such institutions, which are usually governed by a committee elected from the contributors, are designated "Registered Hospitals." Every hospital receiving patients must have a resident medical officer.

Under the law, private patients are consigned to asylum care on the order of a relative, friend, or some person authorized to place the patient under legal restraint, if the document be supported by the certificates of two qualified and registered physicians. In the case of Chancery patients, an order signed by the regularly appointed committee is sufficient authority for the reception of such persons into "any asylum, hospital, licensed house, or other house," without a medical certificate. Paupers are generally sent to an asylum by the order of a justice together with a statement of particulars and a medical certificate. In the event of a pauper patient not being able to be taken before a justice, the clergyman of the parish (in priest's orders) in conjunction with the relieving officer or overseers, as the case may be, can sign the order for his admission into an asylum. But in regard to commitment, it is advisable to quote from an official authority. Secretary Perceval, of the English Lunacy Commission, has described before a parliamentary committee the present system of commitment substantially as follows:

To authorize the confinement of an insane person who has not been found lunatic by inquisition, in any licensed house or hospital, it is required that there should be a document called an order, addressed to the proprietor or superintendent by some person connected in some way with the lunatic, and supported by the certificates of two registered medical practitioners,[1] who must examine the patients separately from each other, and must each sign a certificate in the form prescribed by act of Parliament. Each certificate must state the facts upon which the person certifying grounds his opinion, that the person to be taken in charge is a lunatic within the meaning of the Act, and that he is a proper person to be placed under care and treatment. Those facts since 1853, are obliged to be divided into two categories: first, the facts personally observed by the medical men at the time of the examination; and, secondly, the facts communicated to them, if any, by others, specifying the informant. The medical men certifying must have seen the patient within seven days prior to his admission. The certificates are good for seven days only. The order must have been made within a month prior to admission, and since 1862 the order cannot be made by any person who has not seen the patient within a month. The person who signs the order does it entirely upon his own responsibility; he must state some circumstances in connection with the patient,

---

[1] Under special circumstances a private patient may be confined upon the certificate of one qualified practitioner; but unless two similar certificates be furnished within three days of admission, the patient must then be discharged.

and he must append a statement of particulars giving name, sex, age, and previous abode of the person, which he must also sign; and the statement must contain the name and address of some relation, if possible, to whom notice may be sent in the event of death. No person can sign a certificate or order for the reception of any private or other patient into a licensed house, if he receives any percentage on the payment to be made for his admission; and no medical man can sign the certificate for the reception of such a patient into a house of which he or his father, brother, son, partner, or assistant are the proprietors; nor can his father, son, brother, partner, or assistant sign the same reception order. There is no regulation with regard to the position in which the person who signs the order stands towards the alleged lunatic. All the Act says is, that the order must be signed in the form given in the schedule to the Act. It may be signed by a total stranger if he chooses to take the responsibility [1]; all he has to do is to state : " I, the undersigned, hereby request you to receive A. B., a lunatic (or an idiot, or a person of unsound mind,) as a patient into your house; subjoined is a statement respecting the said A. B." And then there is the name of the person signing, and there is his occupation and place of abode, and his degree of relationship (if any) or other circumstance of connection with the patient. The order is not countersigned by any authority. There is no provision, as there ought, in my opinion, to be, making it the duty of some one to inquire of the person signing the order what reasons he has for signing, or if there are any persons more nearly related to the alleged lunatic than himself. The sole safeguard is the responsibility incurred by the person signing the order, which responsibility amounts to the risk of a civil action on the part of any person who considers himself aggrieved. The person taking charge of the alleged lunatic is required within twenty-four hours to send notice of admission of the patient to the Commissioners in Lunacy, and if he be the proprietor or superintendent of a licensed house in the provinces, he must also send notice to the clerk of the visiting magistrates. He sends that notice of admission accompanied by copies of the

[1] Some measure of responsibility attaches also to the certifying medical man. In the case of Hall v. Semple, decided in 1862, the plaintiff obtained a verdict for £150 damages against the defendant, a medical man who " negligently and culpably " signed a certificate for admission to a private asylum.

order and of the two certificates under which the patient was admitted, and he must at the same time make an entry in his admission-book of the patient's reception ; then after two days, and within seven clear days from the reception of the patient, the medical attendant, whoever he may be (either the medical superintendent of the hospital, or asylum, or the proprietor of the licensed house himself, if a medical man), but in any case the person who has the medical charge of the patients in the house or asylum, must send up to the Commissioners in Lunacy what is really a third certificate, stating what the patient is as to his mental and bodily condition. It is filled up by different medical superintendents with more or less particularity ; some of them fill up the form very fully and carefully, and give a great deal of information with reference to the patient ; some fill up the form by stating that the patient is laboring under "mania" or "dementia" as the case may be. The words are : " I have this day seen and examined the patient mentioned in the above notice," that is the notice of admission, " and hereby certify that with respect to mental state he (or she) " is so and so, " and that with respect to bodily health and condition, he (or she)" is so and so.

With regard to the amendment of certificates or orders, which may not appear to the commissioners to be in proper form, the 11th section of the Act 16 and 17, Vict. reads as follows : " If after the reception of any lunatic it appear that the order or medical certificate, or (if more than one) both or either of the medical certificates, upon which he was received, is or are in any respect incorrect or defective, such order and medical certificate or certificates may be amended by the person signing the same at any time within fourteen days after the reception of such lunatic : provided, nevertheless, that no such amendment shall have any force or effect unless the same shall receive the sanction of one or more of the commissioners." That was extended a little further in 1862 by the 27th section of the 25 and 26 Vict. c. 111, which said : " Where any medical certificate upon which a patient has been received into any asylum, registered hospital, licensed or other house, or either of such certificates is deemed by the commissioners incorrect or defective, and the same are or is not duly amended to their satisfaction within fourteen days after the reception," then the commissioners may make an order for his discharge. A superintendent or proprietor of a house would be safe in taking a patient if upon the face of the documents there was no very gross and palpable omission, such as no facts stated to have been observed by the medical man himself, or a false signature to the order.

A relieving officer or overseer is bound, under heavy penalties, to remove every pauper lunatic, or every lunatic not under proper care and control, whether a pauper lunatic or not, whom he finds or has notice of in his district. That is under the provisions of the 16 and 17 Vict. c. 97, which is the great Act regulating all lunatic asylums and pauper lunacy, sections 67 and 68. Section 67 imposes upon the relieving officer and medical officer of the parish or union the duty of removing pauper lunatics to an asylum, and section 68 refers to lunatics wandering at large and not properly taken care of, or cruelly treated, and not under proper care and control. In the case of paupers the order is the order of the magistrates, not being a mere request, but an actual order to the superintendent of the asylum to take the patient in. It is accompanied by a statement signed by the relieving officer, saying what union the chargeability rests primarily upon, and it is accompanied by a certificate which is in precisely the same form as a private certificate, with the exception that it omits the words relating to a separate examination, because there is only one certificate. Another difference is that the notice of admission is not sent to the commissioners within twenty-four hours, as is the case with a private patient, but within seven days. The medical superintendent of the asylum sends up his statement of condition along with the notice of admission of the pauper patient, which is signed by the clerk of the asylum on some day after two days and within seven days after the admission. The medical certificate lasts till the discharge of a patient.

On the recovery of any private patient, notice must be given to the person who signed the order of admission, or to the one by whom the last payment was made; and if the patient be not removed within fourteen days, notice must then be given to the commissioners, and in the case of a house licensed by justices, to the visitors also. Any patient, not a pauper, may be discharged or removed, whether recovered or not, by the order in writing of the person who signed the order of admission, or in the event of his death or incapacity, by that of some other person authorized to act in his stead, unless the medical officer in charge object to the patient's removal on the ground that he is dangerous and unfit to be at large, in which case the consent of the commissioners or visitors must be obtained for his discharge or removal; though he may be transferred, under the control of an attendant, to some other establishment. If there be no relative or other person qualified to order the removal or discharge, the commissioners may do so.

The commissioners are also empowered to effect the discharge of private patients for two reasons: one for the insufficiency of certificates, if after commitment they are still deemed insufficient; and, secondly, if two of the commissioners visiting at intervals of seven days, and with certain formalities prescribed by the Act, come to the conclusion, that any person held there, to whom their visits are specially directed, is detained, in the terms of the Act, without sufficient cause. "The Commissioners," to use the words of Secretary Perceval, "from their own observations, sometimes think that a patient ought to be allowed to go, and their friends sometimes think that it is not time that they should be discharged. It is more in the case of pauper patients that we hear these complaints than in the case of private patients. A near relation wishes to get the breadwinner of the family out of the asylum, or the husband wants to get his wife back, because he finds it very uncomfortable to be living without her, and he wishes her to be discharged whether she is quite cured or not. These are the kind of complaints we get in much larger number than those relating to the undue detention of private patients."

The committees of visitors entrusted with the local management of county and borough asylums may discharge a patient, whether recovered or not, on the recommendation of the superintendent, or they may do so of their own authority and against the wish of the medical superintendent, provided three of said visitors concur. Generally, however, they act on the recommendation of the medical superintendent. The commissioners have no power of discharge in the case of paupers, other than for insufficiency of certificate. They may, however, recommend discharge for other reasons.

Pauper patients (not dangerous) in licensed houses and hospitals may be discharged therefrom by the Board of Guardians of any parish or union, or an officiating clergyman of any parish not under a Board of Guardians, with one of the overseers thereof, or any two justices of the county or borough in which such last-mentioned parish is situate.

The commissioners frequently advise the transfer of a patient where there is a difference of opinion between them and the superintendent of a hospital or licensed house as to the necessity of confinement. The Home Secretary may discharge absolutely or conditionally any criminal lunatic.

In 1877, an important investigation was made by a select committee of the House of Commons, into the operation of the lunacy laws so far as regards the security afforded by them against violations of personal liberty. The inquiry was particularly directed to what are known as " private asylums "; viz., thirty-nine metropolitan and sixty-one provincial licensed houses kept by their proprietors for profit. Some grave charges were advanced against these private establishments, but with no very definite result. It was shown, that English forms of commitment were susceptible of, and indeed required, amendment: yet it was decided that no abuses of personal liberty were clearly established by the evidence. The committee, in reporting on the evidence, expressed the opinion that frequent and careful visitation is the surest mode of guarding against unduly prolonged detention, and advised the more general adoption of probationary discharge, with full power in the hands of the commissioners to order discharge. It was stated before the committee, that, since 1845, there had not been more than ten cases of discharge effected by the commissioners under the 76th section of 8 and 9, Vict. c. 100, which provides, that, if a patient is to be discharged, and the person who signed the order is contumaciously unwilling to discharge him, two of the commissioners may examine the patient, and, after two separate visits, with an interval of seven days between the visits, may discharge him from confinement. Although it may be true that only ten discharges were effected in this way, nevertheless, it will be seen from the following testimony of the late Chairman of the Lunacy Commission, before the parliamentary committee just referred to, that the moral effect of this law has made the commissioners instrumental in releasing large numbers:

"In going the round of our different houses we saw a number of patients whom we thought in a fair way of recovery, and who might safely be taken out, or who at least should have a chance of recovering by the benefit of a change, and who should be so treated by their friends. We used, as we do now, to enter observations of that sort in the Patients' Book. The friends see those entries, and in a vast number of instances, nay, almost in all, act upon them. We have been in the habit of writing to the relatives of the patient, or the person who signed the order, using this argument: You must be perfectly aware that there is a sore place between you and the patient. You have signed the order, and you have put him in confinement. It is very necessary that we should, if possible, keep up harmony in families. Will you have the grace to take the person out? If you do not do it, we shall be obliged to resort to our powers of liberation; but we think it is far better that the liberation should come from another quarter, from the friends themselves, and it will perhaps heal a sore place and restore kind feelings. The result has been that scarcely ever any resistance is made ; our advice is attended to, and hundreds and thousands are constantly liberated upon that principle. We gave leave of absence during the year 1876, in the metropolitan licensed houses, to 614 persons ; of these, 131 never came back ; we never inquired after them, and were very glad that they did not return to the asylum. The same thing has been going on in the provincial asylums, both in licensed houses and in the county asylums, and in the hospitals ; and in that way a great number of persons have been set at liberty, and have never returned to confinement. A great many go out on trial, and the trial has proved effective in many cases. Some have come back, some have not come back ; and thus many have regained their freedom."

Notwithstanding the above statement by so distinguished an authority, respecting the favorable operation of the English lunacy laws, it would appear that they do not sufficiently guard against the abuse of unnecessarily prolonged asylum detention. The Commissioners in Lunacy in reporting upon the Colney Hatch asylum in 1879, said :

" During our visit on this occasion the complaints of unnecessary detention are unusually numerous, and very numerous also are the complaints in both divisions that access by the patients to any but the medical staff is practically denied. Of course we cannot affirm the insanity of all the patients ; neither is it our province to consider the discharge of any ; but these complaints are made to us. We earnestly trust that the members of the Committee whose duty it is to visit periodically the wards do make those visits real opportunities for listening to the patients, hearing their grievances, and reporting thereon to their colleagues in view to remedy any cause of complaint. Many patients in each division, especially on the women's side, have told us that they know the chairman by sight and are well acquainted with the medical officers ; but by no means few, who have been in the asylum a long period, and who do not lack intelligence, informed us that they knew not that they were visited at intervals by members of the Committee for inquiry into their cases. Some, indeed, knew that they were periodically counted in the wards, but denied that they had then such opportunity as they desired for complaint.

In another direction we would also suggest whether assistance could not be given to the medical staff. We allude to the discharge of cases 'relieved' only. This must be often a difficult question, and must largely depend upon the amount of care and protection which can be secured for the discharged through their relatives. Looking at the vast number of patients, we think that the Committee should consider whether some system of inquiry could not be organized which might relieve the medical staff of labor in this direction. The medical superintendents have certainly no leisure, and the task of inquiry into the circumstances of a patient if discharged as relieved might possibly be undertaken by members of the Committee in rotation, assisted by a clerk. The result would doubtless be a diminution of the crowd detained here, a boon to the partially recovered, and an economy."

The commissioners in again reporting on this asylum in 1886, said:

"We had a great number of complaints as to undue detention, and we referred all the patients to the Committee, telling them that they alone had absolute power of discharge. We learn that the Committee go round once every two months, but some of the patients told us they were denied speech with them; but this we trust is not the case. The patients also complained that the guardians from some of the unions never came to see them, and this complaint we learn is well founded, and have to express a hope that the Committee will urge on the various boards who neglect their duty in this respect the desirability of making visits to their patients at certain times."

As much interest attaches to the practice of liberating patients "on trial," it may be well to give a few details on this point. By the Act of 1845, it was made lawful for the proprietor or superintendent of any licensed house, or of any hospital, with the consent in writing of any two of the commissioners, or in the case of a house licensed by justices, of any two of the visitors of such house, to send or take, under proper control, any patient to any specified place for any definite time for the benefit of his health. The consent of the committing party was then deemed essential, unless for special reasons the commissioners or visitors dispensed with such consent. In 1855, legislation empowered the superintendent of any registered hospital, with the written sanction of any two of his committee of management, to effect the removal of any patient on trial. In 1862, power was conferred on the commissioners, on committees of governors, and on boards of vistors, of their own authority, to permit pauper patients to be absent on trial. The words of the section are as follows :

"Two of the Commissioners, as regards any hospital or any licensed house, and two of the committee of governors of any hospital, and two of the visitors of any licensed house, as regards any licensed house within the jurisdiction of visitors, may of their own authority permit any pauper patient therein to be absent from such hospital or house upon trial for such period as they may think fit, and may make or order to be made an allowance to such pauper not exceeding what would be the charge for him in such hospital or house, which allowance shall be charged for him and be payable as if he were actually in such hospital or house, but shall be paid over to him, or for his benefit, as the said Commissioners or visitors may direct. In case any person so allowed to be absent on trial for any period do not return at the expiration thereof, and a medical certificate as to his state of mind certifying that his detention as a lunatic is no longer necessary be not sent to the proprietor or superintendent of such licensed house or hospital, he may at any time within fourteen days after the expiration of the said period be retaken, as in the case of an escape."

The system of visitation by the Commissioners in Lunacy is as follows : The districts beyond the metropolitan area are arranged into nine circuits, which are divided between six commissioners, alternating half-yearly in their circuits. All of the asylums, licensed houses, hospitals, and single patients are visited by the commissioners every year—some of them twice a year, or oftener. Workhouses with lunatic wards are visited annually. The smaller workhouses containing lunatics are visited once in three years. The metropolitan district is divided into six circuits, visited four times a year by two commissioners, and twice a year by one commissioner, making six visits in the course of the year, besides frequent special visits in special cases. The work of the Lunacy Commission is at present performed by three unpaid and six professional and paid commissioners and a secretary. The latter is assisted by a large staff of permanent clerks. The paid commissioners, three of whom must be physicians and three barristers, receive a salary of £1,500 each, and the secretary is paid £800 a year. Appointed by the Lord Chancellor for chancery cases are two Masters (or judges) in Lunacy, who are salaried at £2,000 each per annum. There are also appointed by the Lord Chancellor three special Visitors receiving each £1,500 per annum. They visit about 1,000 lunatics, of whom one third are in private dwellings and the rest in asylums. The annual expenditures in connection with the Lunacy Commission and the Chancery and Registrar's department for lunatics exceed $200,000. This is exclusive of the cost of visitation by the inspectors of the Local Government Board, who supervise workhouses and outdoor paupers, and are salaried at from £600 to £1,000 each. It is also exclusive of the cost of visitations by local authorities.

The number and distribution of lunatics in England and Wales on the 1st of January, 1887, according to statistics collected by the Lunacy Board, is shown in the table on the following page.

The figures given do not include "249 lunatics so found by inquisition, living in the immediate charge of their committees, and 69 male prisoners who have become insane while undergoing sentences of penal servitude and who are detained in convict prisons." Since 1884, criminal lunatics have been provided for by moneys appropriated by Parliament.

Where Maintained on January 1, 1887.	Private.			Pauper.			Criminal.			Total.
	M.	F.	T.	M.	F.	T.	M.	F.	T.	
In County and Borough Asylums . .	368	425	793	21,587	26,357	47,944	84	21	105	48,842
In Registered Hospitals	1,608	1,489	3,097	103	60	163				3,260
In Licensed Houses:										
Metropolitan .	861	787	1,648	287	507	794				2,442
Provincial. .	691	847	1,538	152	200	352	5		5	1,895
In Naval and Military Hospitals and Royal India Asylum	259	20	279							279
In Criminal Lunatic Asylum (Broadmoor). . . .							392	139	531	531
In Workhouses:										
Ordinary Workhouses. . .				5,217	6,765	11,982				11,982
Metropolitan District Asylums				2,501	2,898	5,399				5,399
Private Single Patients . . .	186	266	452							452
Outdoor Pauper Lunatics. . .				2,308	3,501	5,809				5,809
Total . .	3,973	3,834	7,807	32,155	40,288	72,443	481	160	641	80,891

From this table, it will be seen that there were 11,982 lunatics in "ordinary workhouses." As these include a very large number of the idiot class, the number of insane who are debarred from asylum care is not so large as at first appears. In 1874 the State took a memorable step in granting four shillings a week towards the maintenance of every pauper lunatic placed under asylum care. This grant, designed to improve the treatment of the insane, does not in England or Ireland extend outside of asylums. In Scotland, however, it is applied to the insane departments of poorhouses, provided the accommodation is approved by the Scotch Board of Lunacy. The effect of the grant has been to draw into asylums many chronic cases that were formerly restricted to the inferior accommodations of workhouses.

What are termed in the foregoing table "private single patients" include lunatics of almost every class except destitute paupers. They are kept not only by persons of the medical profession, but by others; in a number of cases by women, some of whom have acquired in asylums or elsewhere a knowledge of mental disease. The price paid for maintenance and care varies from a pound a week up to several hundred pounds a year, according to accommodation, medical and other attendance.

The "outdoor pauper lunatics," numbering 5,809, are those who are harmless and do not require treatment in an asylum, hospital, or licensed house; and instead of being sent to the workhouse are left under the care of relatives or friends and receive a certain allowance from the poor-law guardians for their maintenance.

The relation which outdoor pauper lunatics hold to the Lunacy Board is thus explained by Secretary Perceval: "Pauper lunatics in private dwellings are, as a rule, left to the visitation of the district medical officer. This official and the relieving officer are supposed to be the supervisors of those patients, and the medical officer is bound to visit them once a quarter. He has a fee of half a crown for doing so, and he reports through the clerks of unions to the guardians, and to the Commissioners in Lunacy, the names of all the outdoor lunatic paupers visited by him during the quarter. Those lists come up and are examined at the Commissioners' office, and inquiries are very frequently directed when it appears from the medical man's report that the condition of the patient is unsatisfactory. The medical officer is obliged to put a certificate at the bottom of his return to the effect that he has seen and examined all the above-named patients." The commissioners have no direct control over pauper patients boarded out; they can only advise respecting their proper disposal and care.

It will be seen by the table that 5,399 lunatics are in the "Metropolitan District Asylums," of which there are three, namely, Caterham, Leavesden, and Darenth. The first two are exclusively for the chronic insane, and at the last are schools for idiot children. These asylums are under the control of the Metropolitan Asylums Board, which consists of rate-paying managers, not more than one third of whom are appointed by the Local Government Board. The remainder are elected by the guardians of the poor in the metropolis. The Metropolitan Board was designed to bring about a better classification of the dependent classes by removing the insane from overcrowded and ill-adapted workhouses to specially constructed institutions in the country, with farms adjoining, and with medical and other supervision suited to their requirements. This consolidation of authorities into one central board has furnished a striking illustration of the power of combination; for by this means has been accomplished, in regard to the chronic insane of London, a great and humane work, which could never have been performed singly by any of the local bodies now represented on the Board.

The three "Metropolitan District Asylums" are technically regarded as workhouses within the meaning of the Act of Parliament, and receive only incurable cases, or those not dangerous to themselves or others. Like ordinary workhouses, they come more under the control of the Local Government Board (formerly the Poor Law Board) than under the Commissioners in Lunacy. The latter may make recommendations, but the Local Government Board directs, in such establishments.

By an Act passed in 1862, the committees of visitors having the charge of county and borough asylums, and the boards of guardians of workhouses, with the consent of the Lunacy Commissioners and the President of the Poor Law Board, were empowered to arrange for the reception into workhouses of a limited number of chronic lunatics. This power was intended to meet the wants of localities where there was not sufficient asylum accommodation for " recent and curable cases," and where the workhouse could be certified to have the requisite accommodation for chronic cases ; such accommodation to include a liberal dietary, ample means of outdoor exercise, medical visitation, paid assistants and proper nurses.

A further measure, passed in 1868 with similar intent, provided that any chronic lunatic transferred from an asylum to a workhouse should continue a patient on the books of the asylum " for and in respect of all the provisions in the Lunacy Acts, so far as they relate to lunatics removed to asylums." But, according to Mr. Henley, one of the Local Government Board's inspectors, this attempt to relieve overcrowded asylums has proved a failure, for the following, among other reasons : A patient under this law, though transferred to a workhouse, would remain an asylum patient subject to visitation by the asylum superintendent, who might give orders, and thus introduce a sort of dual management within the workhouse, imperilling both harmony and efficiency. The parties concerned in the new arrangement embraced four different and overlapping authorities ; viz., the Commissioners in Lunacy, the Local Government Board, the Board of Guardians, and the Committee of Visitors. The commissioners very properly insisted on standard requirements for the insane, and the presence of so many different elements led to the practical abandonment of the measure.

The law prohibits the retention in workhouses of any pauper lunatic who is dangerous or who requires restraint. The Local Government Board, as the leading central authority of the State in poor-law matters, is the supreme authority in the government of workhouses. The Commissioners in Lunacy, in the exercise of their powers of visitation, may, however, order any lunatic to be received into any borough or county asylum, or into any registered hospital or licensed house, if it appears to them that the case is not a proper one for the workhouse. The appeal of the guardians against such order, if made at all, must be made to the Home Secretary. Reports of all workhouse visitations, giving details of the condition of the inmates, their dietary, general treatment, etc., are submitted to the Local Government Board, the Lunacy Commissioners making any recommendations they see fit.

It will be seen from the foregoing table that sixty per cent of all the lunatics of England and Wales under cognizance of the Lunacy Board are provided for in county and borough asylums, the average weekly per-capita cost of maintenance in which, during the year 1886, was as follows:

	$s.$[1]	$d.$
County Asylums	8	$7\frac{1}{4}$
Borough Asylums	9	$7\frac{1}{2}$

These averages include provisions, medicines, clothing, salaries and wages, furniture and bedding, and household requisites; and in some of the borough asylums are also included ordinary repairs of buildings, which is the principal reason of the higher average cost in these institutions.

The average annual increase among pauper lunatics during the past twenty years has been 1,470, the increase being

[1] One shilling sterling is equivalent to about 24 cents in United States currency.

greater than the proportionate increase in the population. It is to be noted, however, that this increase cannot be attributed to a greater proportion of paupers; for statistics show that on the 1st of January, 1867, of the population, 4.44 per cent were paupers, and on the 1st of January, 1887, only 2.91 per cent were paupers. In the opinion of some authorities, this increase in the number of pauper lunatics is not so much due to an increase of fresh insanity as to the annual accumulation of patients under treatment which favors longevity.

The subjoined statistics, from official returns, serve to illustrate the proportion of lunatics and paupers to the population in the years 1867 and 1887.

Year.	Lunatics of All Classes.	Pauper Lunatics.	Paupers,[1] (Sane and Insane.)	Population.
1867	49,086	43,031	963,200	21,677,525
1887	80,891	72,443	822,215	28,247,151

It will be seen from the above figures that, of the lunatics, there was 1 to every 441 of the population in 1867, and 1 to every 349 in 1887; and of the pauper lunatics there was 1 to 504 of the population in 1867, and 1 to 389 in 1887.

Having sketched the salient points of the English lunacy system, and shown the classification and distribution of the insane, also the cost of their maintenance, descriptions of some of the English asylums will now be given, attention being directed, with one exception—the York Retreat—to those designed for the pauper and indigent class. The attempt has not been made to describe fully these institutions, but simply to delineate such features as would indicate their general character.

[1] The English statute defines the term "pauper" to mean a person "maintained wholly or in part by, or chargeable to any union, parish, county, or borough."

COLNEY HATCH ASYLUM.

This is one of three pauper asylums under the control of the justices of Middlesex—a county having, according to the last census, 2,929,678 inhabitants and 80,109 paupers in a total area of 181,317 acres. In proportion to size, this county has not only the largest population, but it is also the wealthiest portion of the United Kingdom, yielding a poor-rate of upwards of eleven million dollars per annum. In its pauper population there is a larger percentage of lunatics than in any other county—one in seven of its rate-aided poor being insane.

The asylum is situated about six and a half miles north of the city of London, and it partially meets the wants of that part of the metropolis lying north of the Thames. The building is a large, inexpensive structure of brick, with slate-roof and stone mouldings, having a main or central division of two and a half stories, and wings of two stories each to right and left. From the porter's lodge at the outer gate, the approach is along a gravelled path leading through pleasant grounds. The estate comprises 160 acres, including lawns, meadows, and gardens. The encircling walls are hidden by evergreens. The asylum was built in 1851, and is designed for acute and, presumably, curable cases, including the suicidal and dangerous. It accommodates upwards of 900 men and 1,300 women.

The Committee of Visitors who manage the asylum consists of eighteen members. The administration of this institution is somewhat anomalous, there being one superintendent over the men's and another over the women's department, thus constituting two entirely separate heads. The respective administrative quarters of these two heads are situated nearly at the extreme ends of the extensive range of buildings.

Entering the portal on the men's side, the visitor finds his attention drawn to a tablet recording the humane act of a popular sovereign in these words: "Her Most Gracious Majesty Queen Victoria (whom God preserve) was pleased, in her charity, on the 8th of May, 1849, to found the Victoria Fund for the benefit of poor patients leaving this asylum." The scrupulously clean hall, and stone stairs scoured with white sand, next attract attention. The interior walls are not plastered, but thickly coated with paint of a pleasing tint; those of the corridors and wards are finished in two colors, the dado being darker than the upper portion, which is relieved by an ornamental border near the ceiling. The various sections of the buildings are connected by long corridors.

There are 267 single sleeping-rooms. About 600 patients sleep in associated dormitories greatly varying in size. Most of the bedsteads are of iron pipe and supported on castors. Each is furnished with a hair mattress, and there is a plentiful supply of blankets. Many have canvas bottoms, which, by means of screws and loops, can be tightened, or readily taken off for cleaning. On the walls were pictures and appropriate mottoes, while on the floors strips of bright carpet gave a look of comfort to the apartments. In some of the dormitories for the more quiet class were large mirrors.

One of the long sitting-rooms may be described as representative of the others. It had tinted walls ornamented with numerous pictures, and contained cushioned seats, arm-chairs, and lounges. On the window-sills were flowering plants; ingeniously arranged aviaries were also noted, the birds in which enlivened the room with song, and, at the same time, amused the patients. It being a wintry day, with frost and snow, the inmates seemed greatly to enjoy the cheerful open fires, round which many were comfortably seated. Bagatelle boards, chess, draughts, and other amusements further diverted attention from the dreariness without. The smaller sitting-rooms for the quiet and convalescent class in what is termed the day portion of the ward are better furnished than the others. Here were observed easy lounges, pianos, billiard-tables, books on centre tables, carpeted floors, frosted globes to the gas-jets, and a profusion of ornamental articles. The women's side corresponds to the men's in its general characteristics, except that the number of single rooms is much less.

The inmates being of the pauper class, they are under a uniform system of care and treatment. In the opinion of those in charge, the insane, when given an opportunity, effect their own classification, selecting their associates by a sort of natural affinity. The violent and suicidal are under constant supervision, and are kept separate from others. Patients are often quieted by removal from one ward to another, the change of surroundings having a pacifying effect. In the epileptic ward, the bedsteads are very low, being only about five inches from the floor, and have canvas bottoms.

The women's infirmary, to which is attached a separate dietary kitchen, is furnished with every requisite comfort, including rubber mattresses that may be filled with either air or water for special cases, invalid chairs, and cushioned couches on wheels. The walls of the infirmary were decorated with pictures and flowers. On the floor were rugs and strips of carpet, and the windows were hung with scarlet curtains.

There are three dining-halls for women. Knives with only about an inch of cutting edge and common forks are used at table, as also ordinary white plates and mugs. At the time of my visit, only about thirty-five per cent of the men were employed. The weather being unfavorable for outdoor work, a smaller number than usual were occupied about the grounds. Forty-four per cent of the women were employed. A majority of these were engaged at needle-work and in the laundry. All mending for the institution is done on the premises, as well as the making of dresses,—plain, useful sewing being preferred to fancy work. For the excitable, washing by hand is thought the most suitable form of employment. Some of the washing, however, is done by machinery. The commodious and well-arranged laundry, with dining, sleeping, day, and kitchen accommodations, is, in one sense, a separate department of the institution. It is under the charge of a superintendent and sixteen paid servants. The appliances here are complete and modern. In separate rooms the clothes are folded and ironed. The laundry is supplied with several steam-driven mangles, steam-rotary washing-machines, and wringers. Clothes were drying in the adjoining yard. Many of the women were engaged about the wards; a few were helping in the kitchen; and a small force were picking the filling of mattresses, which is taken out and renovated when necessary.

There is a large recreation or amusement hall provided with piano, orchestra gallery, and stage for dramatic impersonations, where musical and dancing parties are held. The evening entertainments often bring together assemblages of from five hundred to six hundred patients.

The airing-courts, of which there are several, are large spaces, with patches of green lawn, shade trees, shrubbery, flower-beds, and rustic summer-houses,—the whole enclosed by a sunken wall or haw-haw, thus permitting an unobstructed view beyond the grounds.

It was stated that restraint is rarely resorted to, and then only in cases of great destruction of clothing and for surgical reasons. Patients are commonly secluded in padded rooms during paroxysms. Of these rooms, there are eleven on the men's and eight on the women's side. The padded rooms are each lighted by barred windows placed high in the wall, and are lined to a height of eight feet with strong canvas stuffed to a thickness of several inches with cocoa fibre. The floor is covered with a thick mattress of canvas similarly stuffed. It was asserted that, for women, there were no appliances for mechanical restraint.

The attendants average one to eleven patients. The men are uniformly dressed in blue serge. Each wears a leathern belt, to which is attached a bunch of keys, imparting something of a prison-official character to the wearer. The women wear drab gowns, white aprons and caps, and belts to which are likewise attached bunches of keys.

The rooms are warmed generally by open fires, in front of which are strong screens. This form of heating is supplemented in severe weather by hot-water pipes. The coal bins in the women's ward are filled from the corridors through traps, thus avoiding intrusion.

Under the windows, as well as near the ceiling, slides were observed for the purpose of ventilation. The windows generally have iron sashes, with panes 5 x 6 inches. Larger panes are, however, gradually superseding the smaller ones. The upper sashes of the windows are made to open outward. One chimney about two hundred feet high suffices for the domestic part of the institution, including kitchen and laundry, smaller flues leading thereto from the various open fires.

An attractive feature is a spacious chapel, in which the resident chaplain of the Church of England officiates. Roman Catholic and Jewish services are also conducted once a month—sometimes oftener.

The weekly per capita cost of maintenance here during the year ending January 1, 1887, is reported as 9s. 3½d.

The general impression left upon the mind of the writer by the visit was, that the size of the institution is incompatible with its curative aims, and its management not in entire keeping with its object.

HANWELL ASYLUM.

This asylum is famous as having been the place in which, under public auspices, Dr. Conolly, nearly half a century ago, demonstrated the practicability of the non-restraint principle in treating the insane. In visiting an institution which had thus become historic, one is actuated by a strong desire to know how that great reform has stood the test of time. This asylum and Colney Hatch receive the violent, troublesome, and dangerous cases of the various unions or parishes attached to the populous metropolitan county of Middlesex. At Hanwell these constitute about 40 per cent of the inmates. Chronic and harmless cases, as a rule, are sent to Banstead.

The building at Hanwell is situated some seven miles northwestwardly from the city of London. It is a plain brick structure with slate-roof, erected in 1830, and originally designed for from 300 to 400 patients, but since enlarged to accommodate 1,800 or 1,900. Connected with the institution are 100 acres of land bordering the river Brent, and bounded on the other three sides by a brick wall. Eighty acres are under cultivation.

In administration, this asylum presents the same incongruity as Colney Hatch, there being two resident medical superintendents.

The patients, at the time of my visit, numbered 750 men and 1,100 women, classified according to their habits and peculiarities. The more noisy occupied detached wards. The epileptics were in divisions by themselves. At night they are closely watched by special attendants, whose movements are marked by means of electric clocks that are affected by every opening of the door. The bedsteads for epileptics are low and have canvas bottoms. A detached building is used as a hospital for infectious diseases. Patients occupying padded cells are required to be visited at regular intervals during the night. There are on duty at night six male and twelve female attendants. Throughout the institution the attendants average one to every twelve patients. They wear a uniform, that of the female nurses being dark dresses and white caps. The patients are variously attired.

Open fires are used in the day-rooms as well as the corridors. Those in the day-rooms for quiet patients are protected by ordinary fenders, and those in the wards for the refractory by iron guards. A complex system of ventilation at one time in use has been finally set aside, open fires with direct ventilation from the windows, and perforations in the walls, being now relied upon, and the results are more satisfactory. Thermometers are hung in the day-rooms and corridors, and a record is made of the temperature night and morning. In the dormitories, the rule is "fresh air and plenty of blankets."

On the male and female sides, 30 per cent of the patients were provided with single rooms, which was not deemed in excess of requirements for a class of patients many of whom are restless by night. Dormitories on the upper floor should, in Dr. Rayner's opinion, have protected windows; but for those on the ground floor, where the patients are under constant supervision, he preferred the ordinary kind.

The interior walls are frescoed and appropriately decorated—the work of the patients. On the men's side the furniture appeared abundant and comfortable, including lounges, easy-chairs, and pianos. Pictures on the walls, and a variety of decorative objects were observed in the day and other rooms. In some of the apartments the windows had handsome curtains. On the women's side the furnishing was more elaborate. In one of the day-rooms was noticed a jackdaw, a pet of the female inmates, which had been a long time on the premises. Many inexpensive yet pleasing articles for adorning the room had been made by the patients.

The dining-tables are furnished with table-cloths, ordinary plates, knives, forks, and spoons. Comfortable chairs are provided. In the wards for the excitable cases, the blades of the knives have only about an inch of cutting edge. One general kitchen supplies the whole establishment. The tea is infused, not boiled. A bag of tea having been put into a twenty-gallon copper can, boiling water is then turned on; after infusing ten minutes, milk and sugar are added, and in five minutes thereafter the tea is served. The rule is twenty-three ounces of tea to twenty gallons of water. This method of making tea, as may be inferred, is not given because it is thought to have special value as a recipe.

It is the aim of the management to engage every patient at some kind of useful employment. Patients are first placed under observation and tested at an occupation requiring muscular exertion. In cases where there is any doubt about a patient, employment is prescribed requiring muscular force without the agency of tools capable of inflicting injury. Working parties are examined by a physician each morning before they go out to work. The following is an official statement of the nature of the employments within the institution: "Seventy-seven per cent of the men and sixty-six per cent of the women are doing some kind of work. In outdoor labor 280 men are engaged, of whom 183 work with implements, and 97, who are less trustworthy, without tools. Employed in shops and at trades are 153. Of the women, 80 assist in the laundry; 12 help in the kitchen; 224 are employed at needle and fancy work; while 392 are set apart for domestic work in the female wards." It is the opinion of Dr. Rayner, that patients spending much of their time in the open air are less inclined to be suicidal. The theory of treatment here was summed up as follows: "The patient should be well employed and allowed plenty of outdoor exercise." "This," said the Superintendent, "is the great secret of successful treatment in the care of the insane."

Respecting mechanical restraint, Dr. Rayner made the statement, that, in the treatment of the insane, he had never resorted to mechanical restraint except by the use of gloves, and that he used these only in exceptional cases, such as preventing a patient from tearing open a wound. Neither did he use any chemical restraint, which, in his opinion, was worse than mechanical. During the previous twelve months, he thought there had been prescribed in this institution only twenty-four sleeping draughts. He considered them "utterly pernicious." Strong, untearable dresses are used when deemed necessary, and are so made as not to restrain the limbs. There are sixteen padded rooms on the men's side, and about a score on the women's; and in these the floors, as well as the sides, are cushioned. All of these apartments may be readily darkened, the Superintendent believing darkness to be desirable for extremely violent patients. An attendant is not allowed to attempt to handle a violent patient until he can secure the help of a posse of assistants; and the rules are rigid as to attendants reporting each instance of resistance. When a case of this nature is announced, a body of attendants are immediately sent to the spot, and the patient usually succumbs to the inevitable. Sometimes, the expedient is resorted to of removing an excited patient to another part of the house. If the case be decidedly maniacal, the patient is placed in a special apartment in charge of attendants. In the space of twelve months, seclusion had been resorted to among the men on not more than eight occasions, giving a total of twenty-three hours; and among the women ten times, giving a total of twenty-nine hours.

The well-kept airing-courts are quite large. Several of them have about four acres of ground each; in fact, these spaces are of such extent as to somewhat remove the impression of restricted boundaries, especially as the walls and iron fences are ingeniously hidden by trees and shrubbery. It was a wintry day on which my visit was made, and it was noted with satisfaction that the men in the yards were comfortably clad, their clothing including thick woollen capes or short cloaks, that came well down over the shoulders and amply protected the neck, chest, and arms.

There is a large amusement hall, in which the patients hold dancing parties every Monday evening. In the summer, lawn tennis, skittles, quoits, croquet, and other out-door games are freely indulged in. The quoits are made of rope. A high wall built around three sides of a square is provided for a favorite hand-ball game. Picnic parties in summer are also frequent. With pianos, games, books, and papers, ample provision seemed to have been made for the entertainment of the patients. Sunday services are regularly held, and there is morning and evening worship during the week.

The average weekly cost of maintenance here during the year ending January 1, 1887, is given as 9s. 4¼d. per capita. Patients from other counties are charged at the rate of 14s.

The visit, as already indicated, was one of much interest, owing to the historical prominence of the place. Considering the large number of violent and dangerous insane here congregated and the lack of space in some of the departments, it was gratifying to note the comparative quiet and order prevailing throughout the institution. Much of this good result must be ascribed to the large percentage of patients engaged at industrial employments. In the course of this inspection, permission was granted to visit the wards containing the most disturbed cases ; and throughout an extended examination no unusual disturbance met the eye or ear in either wards or airing-courts. Several patients were seen in the refractory wards pacing briskly to and fro, also about the grounds without ; but no instance of struggling with attendants and no application of mechanical restraint came under my observation. There was conclusive evidence that the methods of handling the turbulent elements of which the asylum is made up are founded upon humane principles, and that these conduce in a remarkable degree to good order, cleanliness, and general contentment. Nevertheless, it appeared to me, that, from overfulness, restricted airing-courts and grounds, and the presence of numbers too great for close individual inspection, the asylum could not reach those curative results which the enlightened principles governing it would seem to warrant.

BANSTEAD LUNATIC ASYLUM.

This institution, erected at a cost of £320,000, was opened in 1877. It is structurally designed for the reception of chronic cases, and, like the asylums of Colney Hatch and Hanwell, is directed by a Committee of Visitors appointed by the justices of Middlesex. It is located in the county of Surrey, having been established under the powers of the Act of 1853 already referred to as enabling one county to purchase lands and erect and manage institutions in another county.

The Banstead Asylum occupies a bleak and elevated site on the Surrey Downs, fourteen miles south of London. In connection with the institution are only 118 acres of land, with less than 100 acres under cultivation.

The asylum is built of buff brick, architecturally embellished with those of a darker color. Its general plan is like that of Leavesden and Caterham, there being oblong blocks extending rearward in two parallel ranges from either side of a central administration building. The blocks are connected by one-story brick corridors. The range on the right as one enters the building is for men and the opposite one for women.

Most of the window-sashes are of iron, the panes measuring $5\frac{1}{2} \times 11$ inches, with iron bars outside, corresponding to those of the sash. The asylum was erected to accommodate 1,700 patients; but notwithstanding the small acreage of land, additions have been made to admit of 300 more of the insane.

The resident Medical Superintendent is assisted by four medical officers, also resident. Among the other salaried officials are a clerk to the Committee of Visitors of the asylum, a resident steward and a chaplain, a resident engineer, and a lady organist. The number of attendants in some of the wards is in the proportion of one to twenty patients; the average proportion throughout the asylum is one to fifteen patients. There was an unusual degree of disturbance, possibly owing to the pressure on accommodation. The small airing-courts were so full of patients as to prevent healthful recreation.

Despite the fact that the building was designed only for

the chronic class, many forms of acute insanity were included in the cases received. Little restraint was used except for surgical reasons. Locked gloves with free thumbs were in use, and women were loosely fastened in chairs by towels. The strong dresses worn by some were artfully disguised by fabrics of a bright color; and the ordinary clothing, which seemed comfortable, was also selected with a view to the avoidance of dull or sombre hues.

A large day-room of one of the female wards was overcrowded, the patients were very restless, and presented anything but an agreeable spectacle. Some were crouching or lying on the floor; others, with dishevelled hair, were rushing wildly to and fro.

In the epileptic ward low bedsteads are provided with canvas bottoms, and the patients are under night supervision. Telegraphic communication connects the office of the head physician with every department. Electric tell-tale clocks check the movements of the night attendants, and electric fire-bells are provided and arranged so as to alarm, in case of any outbreak, the fire-brigade, which consists of two experienced firemen assisted by attendants. The food is prepared in a general kitchen, and conveyed to the several blocks. The water supply is obtained from a well sunk to a depth of 300 feet; its capacity being 60,000 gallons per day. The water is hard, and the process of rendering it suitable for asylum use is expensive. The percentage of inmates usefully employed here was quite small. The male patients worked chiefly on the farm and in the garden; a few were occupied at upholstery and other trades, and as helpers in the various wards. The women were engaged principally at needlework, ward-cleaning, and in the laundry.

Amusements and means of entertainment are not neglected. A library is provided, out of which there were in circulation among the patients, at the time of my visit, about 600 volumes. Dramatic entertainments are not unfrequently held, and it was stated that in pleasant weather there are occasional picnic excursions. A band, composed of asylum attendants, furnishes music for the weekly dances and other amusements. About 450 patients attend divine service in the chapel.

The average weekly per-capita cost of maintenance during the year ending January 1, 1887, was 9*s.* 5¾*d.*

In this large and modern asylum there appeared to be a straining after economy, a sacrifice of essentials, and a feeling of discomfort and confinement among the inmates. The writer was surprised to find a comparatively new asylum for the chronic insane, who require a large acreage for outdoor employment, built upon so small a tract of land; and more surprised to see such an institution assuming the functions of a hospital for the cure of acute insanity.

LEAVESDEN METROPOLITAN ASYLUM FOR THE CHRONIC INSANE.

This asylum, opened in 1870, is situated in Hertfordshire, some seventeen miles northwest of London, on a farm of eighty-four acres. It was built to accommodate 1,638 patients, and was subsequently enlarged to receive 2,000. The original outlay on lands, buildings, fittings, and the necessary furniture for its several departments was at the rate of about $430 per bed. Leavesden is one of three institutions for harmless pauper lunatics of the metropolitan asylum district. It is directed by a committee of the Metropolitan Asylums Board, and is visited by the Commissioners in Lunacy. Although in reality an asylum, it is, as already stated, a workhouse within the meaning of the statute, and comes immediately under the supervision of the

Local Government Board, with whom rests the framing of its rules and regulations.

The number of inmates, January 1, 1887, was 885 men and 1,100 women, being an increase of five over the number here at the time of my visit, when the staff and employees were as follows: A medical superintendent with two assistants, chaplain, steward, engineer, clerk and matron, steward's clerk, assistant clerk, assistant matron, head laundress, work mistress, two male head attendants, six first-class married attendants with their wives, three first-class ordinary attendants, first-class night attendant, four second-class night attendants, eighteen second-class ordinary attendants, two female head attendants, nine first-class ordinary female attendants, twenty-six second-class ordinary female attendants, three first-class female night attendants, four second-class female night attendants, hall porter, two first-class laundry maids, seven second-class laundry maids, cook, mess-woman, baker, assistant baker, mess-man, gate porter, three store porters, two scullery men, two outdoor attendants, coal carrier, two laundry men, stoker, earth-closet man, two gas-men, two tailors, assistant tailor, three shoemakers, two upholsterers, grave-digger, cowman, assistant cowman, three gardeners, two carmen, four farm laborers, foreman of works, working engineer, smith, carpenter, painter, ordinary laborer, and plumber.

The asylum is built of buff and red brick. Ivy on the walls adds to the diversity of color and outline, its effect on the chapel being especially pleasing. The visitor is favorably impressed as he approaches the institution, by the grounds, laid out in lawns and beautified by shrubbery and flower-beds; while the plant-houses and conservatories suggest thoughts of summer even in the coldest and most dismal weather. Most of the corridors connect with these conservatories. Each of the blocks accommodates 160 patients, the ground-floor being for day use and the upper one for dormitories. They are warmed and ventilated by open fires, supplemented when necessary by the use of hot-water coils.

The general plan of the sleeping apartments is that of the associated dormitory. There are but twenty-eight single rooms on the male and twenty-four on the female side, the larger proportion being in the infirmaries. One of the dormitories examined is divided into stalls by means of wooden partitions about four feet high. Each stall contains two beds, provided with comfortable mattresses and ample covering. During the day the bedding is so arranged and exposed as to facilitate airing and inspection. Waterproof sheeting, laid beneath the lower blanket, is used when necessary. The bedsteads are of iron, and in most cases have canvas bottoms. The canvas is broadly hemmed, and is so adjusted to the bed-frame as to permit of its being tightened by screws.

Another dormitory was observed to contain eighty beds ranged in four rows, each person being allowed not less than 400 cubic feet of air-space. Lavatories and closets adjoin the dormitories. For the epileptics, who numbered over 200 men and nearly 250 women, the beds are only six inches above the floor. This class are under close night and day supervision, which is checked by electric tell-tale clocks.

A day-room on the ground-floor contained comfortable furniture, including easy-chairs. There were also pictures, baskets of flowers, and song-birds in attractive cages. Five open coal fires in this large and cheerful apartment serve both to warm the room and purify the atmosphere.

In another of the day-rooms a broad belt of bright carpet along the centre of the floor, and rugs in front of the fires, gave a comfortable look to the apartment ; while pictures on the walls, plaster busts, statuary, and other ornaments heightened the pleasing effect. In the windows were plants and flowers in rare profusion. Each of the day-rooms in the female ward is provided with a piano. On the men's side the day-rooms are likewise comfortably furnished and pleasingly decorated.

The interior walls here, as in some other pauper asylums in England, are not plastered, but are painted on the brick and decorated with stencil-work. The upper portion of the wall is generally covered with silicate, and the dado is painted a darker shade. This form of decoration extends to the corridors.

Some of the windows have double sashes, with panes each 16 x 21 inches. The sashes can be raised and lowered a few inches for ventilation. Most of the windows have iron sashes with panes $6\frac{1}{4}$ x 18 inches. They are so made as to be opened six inches at the top. In a block recently erected, the windows have two sashes provided with weights, and there are outside iron bars.

The bath-room is lined with glazed brick, and has a floor of figured tile. Adjoining is a comfortable dressing-room with an open fire-place. Some of the baths are curtained, affording greater privacy. The arrangements for bathing are quite elaborate. The following is a list of the various baths : Cold spray, hot spray, cold wave, hot wave, cold douche, hot douche, spinal douche, hot shower, cold shower, hot needle, cold needle. A Turkish bath is used daily by either male or female patients. Dr. Case, the Superintendent, attaches considerable importance to the hydropathic treatment, particularly for excited patients and for those who do not sleep well.

Food is supplied to the whole establishment from one

general well-arranged kitchen. Cleanliness and order were as conspicuous here as in the other departments of the institution. The walls were tinted, the deal tables well scoured, the tinware brightly polished, and the crockery neatly arranged on shelves. Gas and steam appliances are used for cooking, and the food is conveyed to the wards on three-wheeled trolleys. On the day of my visit there were huge meat pies for dinner, the top of each being encrusted with the number of the ward for which it was destined. The occupants of the female infirmary ward partook of a wholesome and well-cooked dinner at a central table, amply supplied with table furniture, including plates, mugs, salt-cellars, pepper-boxes, and knives with a cutting edge of about two inches.

The inmates were variously employed. In the laundry were seen fifty-five women at work under nine laundry maids, and fifteen men doing the heavier work under the supervision of two paid male attendants. The vegetable garden is extensive. The supply of water is obtained from a deep artesian well on the grounds, that yields 60,000 gallons per day. Sewage is utilized by distribution on the land.

Among the means of restraint are leather gloves. Strong dresses, laced down the back, are also used, and occasionally feeble patients are tied in chairs. On both the male and female sides there are several of what are termed "half-padded" rooms, and on each side one entirely padded. It was said that excited patients were sometimes placed in darkness or treated medicinally. The violent insane are subject to night supervision. There was no case of seclusion or restraint within the institution at the time of my visit.

The recreation hall for females is prettily decorated with light fresco-work. It contained several pieces of statuary and ornamental vases. In the windows were flowers in hanging-baskets and song-birds in cages. Four open fires protected by screens dispelled the chill of winter. Among the other noticeable articles of furnishing were a piano, comfortable chairs, and cushioned lounges. This spacious apartment, fitted up with a stage and other appointments, was used as a workroom as well as an entertainment hall. The weekly dances here were said to be attended by 120 men and 140 women, while at the fortnightly special entertainments the attendance was computed at 200 men and 240 women. The annual costume ball, for which dresses are made on the premises, is a source of much amusement; also the burlesque and dramatic entertainments held from time to time. In summer, cricket and other outdoor games are indulged in. Walking parties go out in favorable weather, under the charge of attendants. Within the eleven months preceding my visit there had been sixty-two walking parties of men, numbering twenty-five each, with two attendants, and fifty parties of women, numbering fifty each, with four attendants. The dress of the patients was warm and sufficient. Many of the women were seen wearing comfortable scarlet woollen shawls. Some of the men were observed in the airing-court on a raw wintry morning wearing over their shoulders deep, heavy capes with thick collars, and evidently enjoying their usual exercise. A few, as they walked about, were contentedly smoking their pipes.

The chapel of the institution accommodates six hundred persons. The services are generally attended by some four hundred patients. The men and women sit on opposite sides of the building, and are about equally represented at the meetings. Besides daily morning and evening prayers, there is a special weekly service for epileptics.

A library of entertaining literature, in charge of the chaplain, is used by both patients and attendants, and is maintained by a small annual grant-allowed by the committee of management. Some years the circulation reaches nearly or quite 10,000 volumes. Considering the extensive use of the library, and the destructive tendency of many of the patients, it seemed remarkable that so few of the books were injured. Among other evidences of humane and considerate treatment, was the great attention paid to the clothing and personal cleanliness of the patients.

The average weekly cost of maintenance for the half year ending September 26, 1887, was 7s. 5d. per capita.

CATERHAM METROPOLITAN ASYLUM FOR THE CHRONIC INSANE.

Like Leavesden, this institution is exclusively a receptacle for pauper patients, designed to relieve ordinary asylums of chronic cases, and to provide better care than the poorhouse affords. It is managed by an unpaid committee of the Metropolitan Asylums Board, and comes mainly under the supervision of the Local Government Board, though subject also to visitation by the Commissioners in Lunacy.

The institution is pleasantly situated in the county of Surrey, about fifteen miles south of London, upon an elevated site commanding a fine prospect of the surrounding country. Connected with it originally were but seventy-two acres, and it was found necessary to purchase additional land to meet the wants of the institution.

The asylum was opened in 1870. Most of the buildings have tile-roofs, and are plain three-story structures of buff and red brick. The several sections are connected, as is frequently the case in Great Britain, by inclosed one-story corridors. Each of the blocks contains about 160 patients, except the infirmary, which accommodates not more than ninety. In accordance with modern ideas, additions made to the asylum since it was first opened are on the two-story plan, the upper floor being reserved for dormitories and the lower for day use. The buildings are constructed with a hollow space between the inner and outer bricks of the walls. The number of single rooms is comparatively small. The windows of the new buildings have ordinary sashes and long outside bars. The sashes open a little way at both top and bottom, and have panes measuring 12 x 18 inches. In the older buildings the panes are $6\frac{1}{4}$ inches wide. The Superintendent, Dr. Elliot, expressed the opinion that iron-grated windows are not necessary for the class under his charge, and that tile-roofs are not desirable.

At the time of my visit the patients numbered 2,052, of whom 945 were men and 1,107 women. They were of the harmless class. Notwithstanding their quiet character, the number of attendants appeared to be inadequate. Relatives, according to their means, are required to contribute towards the maintenance of patients. There is, however, but one standard of care. The average weekly per-capita cost of maintenance during the year 1887, inclusive of provisions, clothing, warming, cleansing, lighting, salaries, furniture, repairs, and medicines, was 8s.

In the infirmary for women were thirty patients. The windows had shades and draped curtains. The walls were tinted and frescoed. Pictures in considerable variety, baskets of ferns, bright flowers on the mantels, trailing plants, and birds in cages contributed to the cheerfulness of the surroundings. A strip of carpet extended along the middle of the floor, with lateral strips between the beds. The bedsteads had adjustable canvas stretchers, as at Colney Hatch. The mattresses were of horse-hair, and were made in three sections. The bedding for each patient consisted of linen sheets, woollen blankets, a coverlet, bolster, and pillow. The head-boards were covered with white muslin, on which were Scriptural mottoes worked in red letters by the patients. Open fires, some with locked screen-guards, shed a happy glow throughout the apartments.

For epileptic patients and others requiring close attention the supervision is continuous through the twenty-four hours of each day. Six attendants—three men and three women—are employed nightly for the oversight of this class, which, at the time of my visit, numbered about 500 patients. No mechanical restraint was in use. It was stated that, during the preceding year, it had not been found necessary to resort to seclusion, nor had any form of restraint been applied. In that time, no serious accident had occurred to any of the asylum patients. There are two padded rooms, one on each side of the asylum. These are rarely brought into requisition. Arm-chairs with wooden bars in front are used for helpless patients. There are no restraining chairs in the ordinary sense of the term.

The day apartments seemed tidy and well-furnished, although not so elaborately fitted up as the infirmary; yet here also were observed flowers and many objects suggestive of home life. Pianos form a feature of almost every day-room. In the upper dormitories pictures were noticed, and flowers in the curtained windows. Four hundred cubic feet of air-space is allotted to each patient in the associated dormitories, and seven hundred cubic feet to each in the single rooms. Hoods over the central gas-jets are so constructed as to assist ventilation. There are ventilating flues in the walls, and the four open fires in each ward, or twelve to each building, greatly aid in purifying the atmosphere. In the recreation hall hot-water pipes are used. Water is obtained from an artesian well. In addition to the ordinary bathing arrangements, there is a Turkish bath, and, on the men's side, a swimming bath measuring 20 x 30 feet. The food is prepared in one general kitchen and carried to the several wards. The inmates are employed to the extent of their ability—some at very light tasks.

The officials wore a uniform of chocolate color, conspicuous on which were the brass buttons belonging to the livery of the asylum. The women attendants wore dresses of a dark color. Every attendant wore a leathern belt to which was attached a bundle of keys. The patients were variously attired, the Superintendent disapproving of uniform dresses. The women wore shawls of bright colors, in the selection of which the taste of each individual was gratified as much as possible. The rule is here adopted of substituting, upon admission, clothes furnished by the institution for those belonging to the patient. If the dress originally worn is claimed by the committing guardian, it is returned ; otherwise it is given to the patient on his discharge, or, in the event of death, disposed of as the management may direct.

An attractive programme of amusements is regularly carried out. A band of fourteen players drawn from the staff of attendants contributes to the musical entertainment of the patients at stated weekly gatherings. The large amusement hall presented quite a gala appearance, being decorated with flowers and flags in great variety. The fittings comprised a stage and other appurtenances, and, altogether, the apartment seemed well adapted to secure the purposes for which it was designed. Theatrical performances are frequently held. Scarcely second in importance as a means of entertainment is the trained singing-class, which holds periodical meetings. Patients of a literary turn are gratified by the use of a library under the charge of the chaplain, while those interested in the current news of the day may direct their attention to the fourteen daily newspapers placed regularly on the tables of the reading-room. Various outdoor sports are indulged in, including cricket, football, and bowling, for which the grounds are specially adapted. Indoor games are also provided. In summer great satisfaction is derived from walking parties organized under the charge of attendants.

Chapel accommodation for about six hundred persons is provided in a tasteful building having suitable interior decorations. The chaplain holds daily morning and evening services, which are attended by between eighty and one hundred patients. At the Sunday morning service the attendance usually numbers about 350.

The system of accounting is comprehensive, showing, since the opening of the institution, the per-capita cost of maintenance, daily averages, price of provisions and other necessaries, also salaries and cost of medicines. The original outlay on lands, buildings, fittings, and furniture for the accommodation of 1,672 patients, is stated to have been £147,000, or at the rate of about $425 per bed.

The patients admitted here are, generally speaking, of the most enfeebled class. As might be expected in an institution specially designed for infirm and debilitated chronic cases, the mortality is great.

Within and without the asylum were evidences of order and cleanliness, among which were the properly arranged kitchen, the well-aired dormitories, the scoured stone steps, and neatly trimmed lawns. Although objections may be made to the unwieldy size of this institution, and the small proportion of attendants to patients, yet, as an attempt to bridge the chasm between ordinary asylums and ordinary workhouses by providing humane care for the chronic insane, Caterham must be pronounced encouragingly successful.

SUSSEX COUNTY LUNATIC ASYLUM—HAYWARDS HEATH.

This asylum opened in 1859, is under the control of the county justices, who appoint a Committee of Visitors to manage its affairs, and, like all similar institutions, is under the supervision of the Commissioners in Lunacy. This county had, according to the last census, a population of 404,027. Its estimated number of paupers January 1, 1887, was 19,175, of whom a little over seven per cent were lunatics.

The institution is situated at Haywards Heath, thirty-seven miles south of London. The buildings, of buff and red brick, are approached through a broad, well-kept roadway, having on either side terraced lawns, spacious cricket and recreation grounds, shrubbery, and flower-beds. The estate contains between two and three hundred acres, comprising profitable farm and garden grounds.

The visitor is here received by a porter in blue and scarlet uniform, who leads the way to the office of the resident Superintendent. The broad flight of stone steps was cleanly scoured, and banked on either side with a profusion of potted flowers and ferns. At the time of my visit there were upwards of 800 inmates, the larger proportion of whom were women. Both acute and chronic cases are admitted. Nearly all of the patients are maintained by the several unions or parishes of the county; a few are charged to outlying districts, or are paid for by their friends.

The asylum is mainly of three stories. The walls are constructed with an air chamber between the inner and outer bricks. In most cases the interior is not plastered, but painted directly on the brick. Plastered walls are, however, preferred. The stairs of white stone are of easy ascent. The windows have iron sashes, and are uniform throughout the building. The panes measure 5 x 10 inches. Two sections of the sash, containing two panes each, open outward.

About two thirds of the patients sleep in associated dormitories which have from three to twenty-five beds each. One of those on the women's side contained nine beds and as many comfortable chairs, besides night-chairs. Strips of carpet were on the floor, pictures were hung on the tinted walls, and the windows had shades. There were also mirrors, and racks for the towels, and the dressing-table had a fancy cover. The bedsteads were of polished wood. The bed furnishing included linen sheets, blankets, a hair mattress, and a pillow to each bed. On each of the white counterpanes was a fanciful monogram embroidered in bright colors. The furnishing of the single rooms was similar to that of the associated dormitories. Iron bedsteads were gradually taking the place of wooden ones. During the night the sleeping apartments are visited every hour. They are heated by large stoves, and thermometers are hung on the walls to register the temperature.

A ward for a better class of female patients was quite attractive. The floors were carpeted, the windows had shades, as well as curtains. The walls were papered, and on them were hung numerous pictures; while the general furnishing included cushioned invalid chairs, a piano, a small library, and bagatelle boards. Birds and plants aided in giving a cheerful and attractive aspect to the room. Outside, on one of the window-sills, were observed several pigeons; on another sat an old, contented tabby—all of them special pets and favorites. The infirmary ward for women is a pleasant apartment containing thirty-eight beds.

The two suicidal and epileptic dormitories—one for men

and the other for women—contain each fifty beds. Those for suicidal cases are so arranged as to bring all the inmates under view at once. Two nurses remain here overnight and make a record of every incident that occurs. The single rooms for epileptics are padded, the floor padding being in three sections. Shutters are provided for darkening the rooms when necessary. The visits of the perambulating night-watch are checked by a tell-tale clock.

The dormitories for men are similar to those for women, except that the decorations are not quite so elaborate. In one of those for epileptic patients were noticed several deep bedsteads. There are four attendants here to forty patients. The detached hospital for men is under the care of a married attendant and his wife.

Only one corridor throughout the entire building is warmed by hot-water pipes, stoves and open fire-places being the usual means of heating. The lighting is by gas. Thorough supervision renders the protection of gas-jets needless.

The patients of both sexes are variously attired. The female attendants wear brown dresses, black aprons, white collars, and frilled caps with green ribbon. The male attendants have dark-blue frock-coats with brass buttons, brown trousers, and caps with leather fronts. The wages of day attendants begin at £26 a year and rise to £40 for men; those of women begin at £16 and rise to £30. In 1883, when the housekeeper, who had been twenty-five years in service, resigned on account of failing health, a pension of £100 a year was secured for her.

One general kitchen supplies food for the whole establishment. Three meals per day are allowed; viz., breakfast at 8 A.M., dinner at 1 P.M., and supper at 6 P.M., the sexes dining in different rooms. In a large apartment with arched roof and tinted walls the women were seen at dinner. They were seated at covered deal tables, at each end of which a "charge-nurse" presided. Crockery, and blunt knives and forks of the pattern common to these institutions, were in use. Each patient receives half a pint of beer at dinner, unless it is withheld for special reasons. The diet, which is fairly liberal, is varied daily, and includes bacon, roast meat, pickled pork, Irish stew, mutton broth, suet pudding, meat pudding, bread and cheese, rice, and fresh fruit in its season. Extras are given to patients as inducements to work. These include tobacco and snuff.

Water is supplied from a well which yields fifty gallons per day for each inmate. Fire-hose is at all times readily accessible in the various departments. The sewage is used to fertilize the land connected with the institution. Waste water and sewage run direct to the sewage farm without intercepting catch-pits or tanks. At each angle in the sewers there is an opening for ventilation and inspection. A system of detached closets has been introduced here.

What is designated as "wet and dry packing" is practised for medical reasons, and is the only form of mechanical restraint, if such it may be called, in use in this asylum. This treatment the Commissioners in Lunacy regard as restraint, and require that it shall be so reported. Opinions differ as to its value from a curative point of view. Locked coats, dresses, and stockings are occasionally used to prevent a patient from scratching or otherwise injuring himself; but his arms, hands, and legs remain free. Seclusion is sometimes resorted to.

In this asylum there is a thorough industrial system, and it is considered of the utmost importance to have a variety of congenial employments. For the men there are sixteen and for the women six different kinds of occupation. These include the following: Farm and garden work, shoemaking, tailoring, carpentering, bricklaying, painting, upholstering, mat-making, basket-making, laundry work, and needlework. Patients are also employed in the kitchen, the dining-hall, the store-rooms, the bath-rooms, and as ward cleaners, house attendants, bakers, engineers, etc. The furniture, baskets, mats, and mattresses, as well as the clothes, boots, and shoes of the inmates are all made on the premises under the direction of skilled artisans who work with the patients. Seventy-six per cent of the insane are usefully employed.

The following is a complete list of the officers, attendants, servants, and artisans of the establishment: Medical superintendent, senior assistant-medical officer, junior assistant-medical officer, chaplain, clerk to visitors, clerk of asylum, housekeeper, head female attendant, female assistant officer, storekeeper, head male attendant, twenty-three male attendants, twenty-eight female attendants, five male night attendants, five female night attendants, organist and schoolmaster, office clerk, seamstress, head laundry-maid, three laundry-maids, hall porter, man cook, dairy-maid, kitchen-maid, vegetable-room maid, three housemaids, two farm attendants, house attendant, bath attendant, engineer, bailiff, gardener, blacksmith, front ground laborer, tailor, shoemaker, mat-maker, basket-maker, upholsterer, baker, butcher, under garden attendant, carpenter, bricklayer, bricklayer's laborer, painter and glazier, stoker, under stoker, cowman, under cowman, two carters, stockman, and house carter.

During the year ending January 1, 1887, the average weekly per-capita cost of maintenance was 8s. 11¾d. Pauper patients from other counties or boroughs are charged 14s., and all private patients, 16s. per week.

The wards of this asylum are well supplied with books, papers, and periodicals. Eighteen daily and fifty weekly newspapers and thirty-eight monthly publications are provided for the use of the patients. It is customary to expend about two hundred dollars yearly in the purchase of books. Many of the inmates are permitted to occupy a portion of their time in writing and in copying selections from literary works, which are favorite ways of spending time with some of those who are best educated. In winter, the chaplain conducts a school in the large dining-hall, which is attended by about sixty patients of each sex. The three Sunday services in the chapel are attended by about six hundred and fifty of both sexes. In the dining-hall of each division the chaplain reads prayers daily, at which services there are on an average about two hundred and fifty men and three hundred women.

The extensive airing-grounds are surrounded by a haw-haw, so as not to obstruct the view. In fine weather, parties of male and female patients not exceeding eighty go out separately into the country for walking exercise, each party in charge of five or six attendants. For the old and infirm, drives are regularly arranged; and in summer there are picnic excursions to Brighton, the Crystal Palace, and other favorite resorts. For the younger people, cricket and croquet furnish healthful recreation. The large amusement hall, in which weekly dances are held, is fitted up with a stage, and is furnished with a piano and other appurtenances for theatrical entertainments and concerts. A fancyd-ress ball is held during the winter, which is largely attended by patients in costume. An average of three hundred of both sexes resort to the weekly entertainments.

Printed rules of the institution were hung in a conspicuous place. They specify the days on which visits may be made by outsiders, and the length of the visits; prohibit the bringing in of wine, beer, or spirits, and direct that nothing shall be said or done that will depress the patients. Only under exceptional circumstances is a patient allowed to be visited until a month has elapsed from the date of his admission. Friends are communicated with in the event of serious illness. The institution was clean and orderly, and bore evidence of enlightened management and considerate treatment of the class under its care.

BROOKWOOD ASYLUM, NEAR WOKING, SURREY.

Brookwood lunatic asylum, like Haywards Heath, is a type of English county institutions which differ widely from the larger metropolitan receptacles already described. It is one of the three asylums for the county of Surrey, and is designed mainly for pauper patients. This county, which includes part of the metropolis south of the Thames, ranks fourth in England as regards population, containing, according to the last census, 1,441,017 inhabitants. On the 1st of January, 1887, there were in Surrey 42,021 paupers, 11.54 per cent of whom were lunatics.

Like other county asylums, the management of Brookwood is in the hands of an unpaid Committee of Visitors. These country gentlemen, most of whom are men of wealth and influence, serve generally for a long term of years, it being customary to re-elect them, unless they express a wish to no longer bear the responsibilities of the position. The Medical Superintendent controls the selection and retention of his subordinates.

This asylum, situated about twenty-eight miles southwesterly from London, is approached through highly cultivated and elaborately ornamented grounds. Its plain, substantial, brick buildings are, for the most part, three stories high. At the time of my visit, the inmates numbered 421 men and 629 women, including acute and chronic cases.

The interior walls of the older portion of the asylum are unplastered; the newer portion is plastered on the brick, a method which is preferred by the Superintendent. Generally, the walls are tinted, and decorated with stencilled frieze and bordered dado. A pleasing effect is thus produced at a trifling cost, the work being done mostly by the patients.

Dining-rooms, with scrupulously clean sculleries attached, are provided for the several wards. These accommodate from thirty to one hundred patients each. The knives and other table furnishings resemble those used in ordinary housekeeping. In a large dining-room that I entered there were small tables, accommodating eight persons each. Its heating arrangements, though in keeping with modern ideas of ventilation, were somewhat peculiar. These consisted of two chimney-stacks standing near the centre of the room, and so placed as to apportion equally between them the space to be warmed and ventilated. Each stack was built with two open fire-places back to back. The windows had heavy draped curtains, and the appearance of the room was inviting. There are two general kitchens, one for the older portion of the asylum and one for the new section.

Another room worthy of note is that set apart for new arrivals. It was appropriately furnished with lounges, settees, and easy-chairs, and made attractive with pictures and statuary.

The day-rooms for both men and women were well furnished. In one examined, the tables had fancy covers; on the comfortable chairs were tidies made by the patients; on the tables and in the windows were potted plants; and pictures were hung on the walls, which were suitably decorated.

In the dormitories, the bedsteads are generally of iron, and are furnished with straw and hair mattresses, except

where woven-wire bottoms are used, in which case the palliasse is dispensed with. The sheeting is unusually thick and strong. Narrow borders of blue or red stripe distinguish the bed linen of the men from that belonging to the women. In the older building, the number of beds in the dormitories ranges from three to twenty. In the new section the dormitories are much larger, each having as many as forty beds. These apartments are warmed by open fires. In the female epileptic ward were ninety-eight patients—some violent and dangerous. All were under watchful and systematic supervision by night and day. Here, as in some other wards of the asylum, was a piano for the entertainment of the patients, a leading principle in treatment being to engage the mind by every possible means. The open fire was only partially screened, yet it was asserted that no accident from burning had ever occurred in the institution.

The new department, intended to accommodate seventy-five female patients in each of its four wards, was erected at a cost of $484 per bed. The open fires have no guards of any kind. A small opening may be made in the windows by means of a lock and key. Ventilation is further effected by flues and perforations in the walls. The panes, twenty to a window, measure $9\frac{1}{2} \times 16$ inches. Thermometers are hung in the various wards, and a record of the temperature is made at different hours of the day. The aim is to maintain a uniform temperature of about sixty degrees Fahrenheit. The single rooms of the asylum, which are in the proportion of one room to nine patients, are unlike in coloring, while the ornamental work is well adapted to please and interest the patients. The furniture throughout is comfortable.

The bath-room, which is of modern construction, has a dado of white glazed tile. The baths are arranged in stalls.

Before a patient enters the bath it is an imperative rule that the temperature of the water, as well as of the atmosphere, must be ascertained by a thermometer and recorded. The temperature of the water must not be less than eighty-eight nor more than ninety-eight degrees. In the event of a thermometer being broken, the orders are peremptory that the bath be suspended until another is obtained.

About five hundred yards from the main building is a small, conveniently planned cottage-hospital furnished with modern appliances. It accommodates sixteen patients, and is arranged for the use of both sexes.

There is separate provision for Church-of-England and Roman Catholic services, which were said to be regularly and largely attended.

The average weekly per-capita cost of maintenance during the year ending January 1, 1887, was 9s. 6¼d.

In this well-ordered institution there was a remarkable absence of ordinary methods of restraint. It was claimed with great satisfaction by Dr. Brushfield that he had never in his life ordered or sanctioned mechanical restraint of any kind. He admitted having met occasionally with cases in which such treatment might have been beneficial; but the risk of abuse was so great that he had been led to discountenance it entirely. Seclusion had not been practised under his *régime* since 1871. He believed this practice led to the neglect of some of the worst cases; he therefore resolved to depart from it, and he never regretted its abandonment. On being asked what he meant by "mechanical restraint," Dr. Brushfield replied: "The application of any thing that hinders the patient in the free use of any part of his body." When he had a troublesome case he preferred to treat it in the ward. If a patient required seclusion, he would be placed in a room with two or three attendants;

but in no case was a door locked. In fact, the attendant who ventured to lock a door under such circumstances would be at once discharged. Formerly there was no padded room; now there are two, one on each side of the asylum. They are rarely used, being required only for very violent and epileptic cases. Special restraining contrivances, the Doctor thinks, only increase excitement. He had frequently found it advantageous to remove a patient from one ward to another. He did not classify to any extent, and had no refractory ward. Classification, in his opinion, might easily be carried too far. Seclusion he defined ordinarily to mean "keeping a patient in a room with locked doors." In that sense he did not permit seclusion, regarding it as a device for enabling and encouraging subordinate officers and attendants to shirk their duty or abuse the patients. Each patient is weighed on admission and at different times during his stay in the asylum.

The male attendants are attired in blue suits. The female nurses wear dark-green dresses, white aprons, and caps trimmed with ribbon. The attendants throughout the institution are in the proportion of one to ten patients. The wages of a first-class male attendant commence at £32, increasing £1 yearly to a maximum of £40 per annum, with board, lodging, washing, and each year two suits of uniform. Second-class male attendants begin at £28, increasing in the same way to £32 per annum, with board, etc., as stated. First-class female attendants commence with £20 per annum, rising to £26; and second-class at £16, rising to £20, with board and minor perquisites. Since my visit there have been some slight changes in the salaries and wages. The doors of the attendants' rooms have large glass panes and curtains, the latter being drawn aside during the day to admit of ready inspection. The reports of events during the night are minute in detail. The day reports, which must be submitted before 9 P.M., set forth the extent and kind of the employments of the day—if unemployed, they state the cause—; also the number taking recreation, the number in bed and their condition, the number wearing locked boots, strong or special dresses, acts of violence, attempts at suicide, accidents if any, and even the minutest incident.

Connected with the institution are 224 acres of farm and garden land. The cultivation of this affords healthful employment for the patients. As many as eighteen acres are tilled with the spade. The gardener is a skilled nurseryman and florist. The farm lands are superintended by a resident bailiff. In the farm cottages, of which there are five, reside quiet patients who work in the fields. The farm, garden, and dairy are highly profitable. The barren heath which surrounded the institution in its early days has been converted into a garden of fertility. The first step taken was to lay out a nursery for the propagation of plants. This eventually repaid all labor expended upon it, by yielding sufficient shrubbery for the ornamentation of the grounds. Subsequently it not only supplied the needs of the asylum, but proved a source of revenue from the sale of surplus plants. A large conservatory upon which much care is bestowed adds to the attractiveness of the institution. This furnishes throughout the winter a plentiful supply of choice and fragrant flowers for the various rooms and wards of the asylum. Dr. Brushfield considers suitable occupation one of the most important factors in treatment, and remarks: "So long as the number of the employed is double that of the unemployed, I am satisfied."

About 120 patients were daily at work under the direction of the farm bailiff and the gardener, and were about equally divided between the two. Other male patients were occupied as follows: Helpers in wards, 32; in kitchen and vegetable room, 25; in domestic offices, 21; as bricklayer, 1; bakers, 4; upholsterers and polishers, 7; basket-makers, 4; bookbinders, 4; mat-makers, 3; shoe-makers, 8; tailor, 1; painters, 7; plumber, 1; assisting engineer, 5; joiners, 3; and in the laundry, 8. Paid artisans not only oversee, but work with the patients in all the mechanical departments. On the female side upwards of 400 were daily employed. The following list shows the occupations: Helpers in wards, 144; sewing and quilting, 106; mending, 42; laundry work, 37; making men's clothing, 36; in kitchen and dining-hall, 26; knitting and netting, 14; boot and shoe binding, 4; assisting housemaid, 4. Owing to an increasing amount of laundry work, steam machinery had been introduced, although Dr. Brushfield favors washing by hand for the robust and excitable, and ironing for the depressed. Since my visit, a separate foul-linen wash-house has been erected here, with suitable disinfecting and rinsing tanks. The payment of money to patients as a reward for the labor they perform is regarded with favor, and has been attended here, as elsewhere, with good results.

Though a strong advocate of employment as a legitimate means of cure, Dr. Brushfield did not push his theories in this direction so far, to use his own language, as to make the asylum simply "a well-regulated workhouse." He regards every asylum superintendent as " primarily a medical officer of a hospital for the treatment of insanity, and one who looks to the treatment of the patient as his first and most important duty." Founded upon these solid principles,

Brookwood has justly attained a high reputation for thorough as well as economical administration.

While the importance of useful labor is properly recognized, recreation and amusement are by no means neglected. The large recreation hall, accommodating eight hundred, is ornamented with statuary, and amply supplied with the usual appointments of a theatre. A number of patients were seen at work adjusting stage apparatus for a forthcoming entertainment. Fancy-dress balls are held every year. These were spoken of as a source of much amusement. They are generally attended by about five hundred patients and a limited number of the general public. For these interesting occasions ladies of the district supply dress material, which is made up by the patients. At the weekly musical meetings two bands perform under the direction of attendants. Programmes for these gatherings are printed on the premises. Each ward has at least one bagatelle board. There is a well-stocked library to which the patients have access, while the supply of serial and comic literature is abundant.

The deportment of the patients was orderly, and the general aspect of contentment was quite marked. No one was permitted to crouch on the floor or to remain in any other unbecoming position, nor was there any evidence of that want of self-respect which is allowed to remain uncorrected in some institutions. In the discharge of the duties devolving upon them, the attendants appeared equal to their assumed responsibilities. Without architectural display or other extravagance, comfortable and proper provision for the insane seemed to be here attained ; and appropriate as were the fittings and furnishings of this institution, that which impressed me most favorably was the humane spirit pervading its entire administration.

THE WEST RIDING PAUPER LUNATIC ASYLUM—WAKEFIELD.

Yorkshire, comprising in its three divisions, or ridings, an area of 6,067 square miles, is the largest county in Great Britain. In population it ranks next to Middlesex. It had, according to the last census, 2,894,527 inhabitants. The estimated number of paupers in Yorkshire on the 1st of January, 1887, was 75,915. Of these, 7.43 per cent were lunatics. The population of the West Riding, according to the last census, was 2,197,811. The estimated number of paupers in this division on January 1, 1887, was 53,803, of whom 4,247 were lunatics. The pressure for asylum accommodation has been so great in the West Riding that, since my visit, the justices have put in course of erection a new asylum to accommodate 1,310 patients. It is situated in the township of Menstone, and has 325 acres of land, which were purchased at a cost of £20,720, or upwards of $300 per acre. The buildings are estimated to cost £188,992, or about $700 per bed. When this institution is completed there will be three large asylums for the West Riding. The other two divisions of the county have one asylum each.

The asylum at Wakefield, a three-story building, accommodating about 1,400 patients, cannot be regarded as a model structure. It is built of brick, and wears the sombre hue of age, having been opened in 1818. Lawns, flower borders, grass-plots, and carefully tilled vegetable gardens, surround the institution. The estate contains 120 acres of land.

The Superintendent, who occupies a separate residence, had, at the time of my visit, five medical assistants. There were two chief attendants on the men's side, with fifty-four subordinates for day and five for night duty. On the women's side there were two head nurses, with fifty-seven assistants by day and five for night duty. The general average of attendants to patients was one to ten ; in the refractory wards, one to five ; and in the suicidal wards, one to eight. The salaries of male attendants ranged from £30 to £50 a year, in addition to uniforms. This compensation, which is here considered liberal, secures long and reliable service. It was officially stated in 1886 that forty male and twenty-one female attendants had been in the service of the institution over five years, and that, during the previous year, only one attendant had been discharged ; while one, during the same period, was retired on a pension.

As we entered the day-room of the convalescent ward for women, a patient was playing a lively air on the piano. The furnishing included large tables, arm-chairs, a sewing-machine, book-cases, pictures, and a variety of minor ornaments. Two open fires, with rails three feet high in front, warmed and ventilated the apartment. There was considerable variety in the dresses of the women. Ribbon bands, generally green, were worn around the neck and waist. The attendants were neatly attired in dresses of blue serge, with white caps and linen collars; but there was a jarring jingle of keys suspended from their leathern belts.

The dormitory on an upper floor had bay-window projections, within which were wash-stands and toilet ware. Several good-sized mirrors were hung on the walls. The wooden bedsteads are constructed with an inner framework, on which a canvas bottom is stretched. Each bed is supplied with a hair mattress and an abundance of comfortable bedding. The sheets of this department are distinguished by a red border. By means of partitions four feet high the central space is divided into stalls ten feet wide, in which the beds are placed. Each stall has a strip of carpet on the floor and a cushioned seat. The nurses occupy one end of the dormitory, from which the entire apartment can be supervised.

Several patients were seen wearing dresses buttoned behind, and occasionally there was one wearing locked boots. In the epileptic ward the patients are by night constantly under the supervision of a regular watch, whose attendance is checked by tell-tale clocks. Records of temperature are made by attendants three times during the night. On the walls of the suicidal ward were noticed many colored prints, and in front of the open fires screens are used during the night.

The refractory ward on the women's side contained armchairs, wooden settees, upholstered lounges, also rugs and strips of carpet. There were high railings before the fire. The pictures were hung within easy reach. Single rooms here have wooden bedsteads with canvas bottoms, a rug by the side of each bed, and sliding shutters to the windows. The floors were cleanly scoured. To this, as to the other wards, a small scullery, closet, and bath-room are attached. The occupants of this refractory ward numbered forty. Twenty slept in single rooms, eleven in one dormitory, and nine in another. The corridors have tinted walls decorated with pictures.

The general bath-room measures 20 x 50 feet, and contains ten baths ranged on either side of a central passage. Muslin curtains adjusted on a framework of gas-pipe posts and rods are provided, shutting in three sides of the bath. The heating is effected by means of hot air. The windows are of frosted glass. The taps of hot, cold, and waste water, are under the sole control of the attendant, and every thing is kept attractively clean. In an adjoining apartment are the usual appointments of a Turkish bath. The baths include vapor, sitz, overhead, needle, shower, and the ordinary douche.

A building of recent construction, for convalescents, is

two stories high, and accommodates one hundred and twenty male patients. Here, in the first room entered, were seen rows of pigeon-holes for boots and shoes. Adjoining this is a lavatory furnished with the usual appliances, including roller-towels and looking-glasses. The rooms were warmed by open fires in front of which were screens $2\frac{1}{2}$ feet high. Comfortable couches, cushioned seats, rugs, and pictures were included in the furnishing. Five attendants were on duty here by day, but there was no special night watch.

The day-rooms on the men's side contain pianos, billiard and bagatelle tables, arm-chairs, and pictures hung on the papered walls. These rooms are heated by open fires. In connection with the men's infirmary is a small glass-covered conservatory containing a choice collection of exotic and other plants and a small aquarium. The conservatory is furnished with comfortable seats.

In two houses detached from the main building upwards of fifty chronic and harmless insane are kept. Want of accommodation is felt here as in almost every other asylum in England. There are, in all, one hundred single rooms in the institution and on three floors there are thirty strong rooms with flush-panelled doors opening outward, so that they cannot be blocked by the patients. The walls of these rooms are covered with silicate, and the window-panes measure 6 x 10 inches. A room on the female side is padded in sections. It has a bed on the floor, and a double-sash small-paned window protected by a padded shutter adjusted by weights. A dormitory on the first floor of the asylum is allotted to the fire brigade, which is composed of attendants and patients, who are called out every month, without previous notice, for training and inspection. Fire extinguishers were observed about the building.

The women take their meals in their wards, sometimes as

many as one hundred dining together. The general dining-room for men is wide, and measures one hundred feet in length. It has high windows, and is fitted up with a stage at one end and a gallery at the other,—the apartment serving the double purpose of dining and recreation hall. The tables at the time of my inspection were arranged for dinner, with linen table-cloths, crockery, and ordinary knives and forks. Twenty-two sat at each table. Before commencing the meal, all stood and sang grace. The dinner consisted of boiled beef and pork, potatoes, carrots, parsnips, and bread, with beer brewed on the premises. As many as three hundred and sixty dine together in this hall. After dinner they remain chatting and singing until it is time to resume work. Meals are supplied from the general kitchen, in which every thing appeared scrupulously clean.

Over one thousand of the inmates are engaged in some kind of work. About two hundred and fifty are employed in farm, garden, and other outdoor work. A peculiar feature of this institution is its large department for weaving, this being an important industry in the district, and one with which many of the inmates are familiar. There were twenty-nine patients employed in the making of sheeting and other fabrics. Fifteen looms, and a large number of bobbin wheels turned by hand, were ranged on two sides of a long building. This department is under the control of an artisan attendant. In the carpenter's shop, six is about the usual number employed. Tailoring is an important branch of the industrial section, eighteen patients being employed at this trade. Seven of the tailors had served their apprenticeship in the establishment, and the instructor informed me that, in his time, seventeen had learned the trade after becoming inmates. The tailor's workroom is well adapted to its purpose. It is warmed by an open fire and is well lighted. It contained three sewing-machines. The foreman, who had been here more than twenty years, said he had never known a patient to hurt himself or others with the tools used. Making, as well as repairing, is done here. The workers were, at the time of my visit, busy on a batch of strong linen rugs, machine-stitched in the form of small squares, each rug measuring 2 x 2½ yards. These were intended for night covering for patients of destructive tendencies. Cloth of a dark color is used for the patients' jackets, and that of a light color for their trousers. The dark-blue uniforms worn by the attendants are made here. In the shoe-shop, sixteen men were at work making and repairing shoes. Various machines are used for cutting leather, riveting, and for other work. The attendant, who had been here many years, stated that no casualty had occurred within his knowledge from misuse of tools. In the tinware-room two patients were employed, and there was the same number in the plumber's room. Picking horsehair and making mattresses gave employment to thirty-five patients in another department. A machine for renovating the hair of mattresses had been in use for a number of years, and was highly approved by the management. In the book-bindery, in which were all the requisite appliances, three patients were at work, while a like number found employment in the blacksmith's shop.

The laundry is a separate department and is presided over by a head attendant. It is extensive, well-arranged, and furnished throughout with the latest machinery. Here were engaged one hundred women under five attendants, with a few laundry maids. Six male patients in an apartment partitioned off from the female quarters were engaged under an employee in doing the heavier work. In washing the lighter articles a number of hand machines were used.

Of the 1,405 patients here, there were only 181 males and 190 females unemployed. The men were not at work for the following reasons: Sixty-two were too sick or too feeble, 38 were aged and infirm, 11 were too low-spirited, 31 too much excited, 33 had too little mind, 6 were able but unwilling. The women were not employed for the following reasons: Twenty-five were too sick or too feeble, 37 were aged and infirm, 3 were too low-spirited, 62 too much excited, 48 had too little mind, 15 were able but unwilling.

The gas for the establishment is made on the premises. Water from the town is used, and there is also an artesian well on the grounds. Attempts to utilize the sewage on the farm had been unsatisfactory, and it was afterwards discharged into the public sewerage system.

The old adage, "A place for every thing and every thing in its place," applies to the minutest details of this establishment. An illustration of this was seen in the yard, where are a few small out-houses, or stores, with an aperture in the door of each, through which waste articles may be passed. The locked doors are labelled; as, for example, "broken glass," "old tin," "old lead." The contents of these receptacles are regularly removed.

It was said that, in favorable weather, some six hundred patients were permitted to enjoy exercise about and beyond the grounds. About five hundred attend worship every Sunday, the Church-of-England service being the form used. Nearly seven hundred, with about equal numbers of men and women, were recorded as attending the weekly entertainments.

Very little seclusion is here practised, and it was said that mechanical restraint is never resorted to except for surgical reasons. This is corroborated by the Commissioners in Lunacy, who, in a late report on the institution, remark that during the year "the use of mechanical restraint or seclusion has in no instance been found necessary." On the death of a patient a *post-mortem* examination is made, unless, on admission, the friends had filed an objection with the management. The average weekly per-capita cost of maintenance here for the year ending January 1, 1887, is given as 8s. 5⅝d.

THE SOUTH YORKSHIRE PAUPER LUNATIC ASYLUM—WADSLEY.

This asylum, three miles from Sheffield, is a modern institution compared with that of Wakefield for the same division of Yorkshire, having been opened in 1872. It was built for the pauper insane, but at the time of my visit it contained a few paying patients who had been temporarily received. The total number of inmates was 1,160, of whom 660 were women. The buildings are situated on a rising slope overlooking an extensive fertile valley. They are of brick, with stone caps and sills to the doors and windows, and the roofs are of slate. After passing the porter's lodge near the iron entrance gate, the visitor proceeds a quarter of a mile along a gravelled roadway to the asylum buildings. A substantial stone church is an attractive feature of the place. The broad green lawns are beautified with laurel and other evergreens, which are disposed so as not to obstruct pleasing views of the surrounding country.

In the main entrance hall there was a profusion of plants in full bloom, that had been brought from the conservatory and here tastefully arranged. The greenhouse is large, and supplies an abundance of bedding plants for the grounds.

One of the day-rooms on the women's side, which may serve for a general illustration, presented a cheerful appearance. It had a large bay-window and two open fires with rugs in front. The floor was carpeted and the walls were papered. Statuary, pictures, hanging-baskets, wicker and other kinds of easy chairs contributed to the general coziness.

In the men's infirmary the windows have double wooden sashes with panes about 5 x 15 inches. On the floor were broad strips of oil-cloth with bright borders. Easy seats were provided, and a few pictures were hung on the walls. Adjoining this ward is a small conservatory, in which were seats for patients.

The refractory wards, like the others, are suitably furnished, and brightened with pictures and other decorations. In connection with each of these wards are one or two secure single rooms. There were no screens in front of the open fires. There are two padded rooms on the men's and two on the women's side. These have double doors, and the mattresses are placed on the floor. In the institution at the time of my visit there were about eighty male and an equal number of female patients of the epileptic class. They were under the special supervision of attendants by night as well as by day.

In addition to a system of heating by means of hot-water pipes, fire-places or open grates are in every ward. Locked fire-screens are used only in the wards for suicidal and epileptic patients. There are a few movable but no fixed guards to the windows. In one corridor was observed an arrangement to promote the health and comfort of the patients, which is worthy of general adoption. This consisted of a range of cupboards with shelves made of slats, under which are pipes heated by steam. Here are placed the damp garments of the patients after they return from outdoor work. A large lavatory which adjoins this corridor, is provided with roller-towels. These are hung above coils of steam pipes to facilitate drying. Satisfactory arrangements for bathing are connected with each ward.

The men dine in a spacious hall; the women in their wards. On the day of my visit the dinner consisted of roast pork and rice, which seemed to be well cooked and was apparently relished. Table-cloths, crockery, and ordinary knives and forks were in use. The kitchen was clean and attractive. Twelve patients were here engaged paring potatoes. In the bakery, aërated bread is made. The dough is kneaded by steam power. Adjoining the bakery is the brewery containing three large vats and all the requisites for brewing. Here is made the beer which is served out to the patients at the daily rate of one pint to workers and half a pint to non-workers. Since my visit, the dietary has been changed so that only patients working in the hay-field and the attendants composing the band are furnished with beer. Water is substituted for it at the patients' mid-day meal. The attendants generally are allowed 1s. 6d. per week to men and 1s. per week to women instead of this beverage.

The laundry is a separate and quite extensive establishment connected with the main building by a corridor. The clothes are successively passed into washing, wringing, ironing, and sorting rooms. In order to furnish desirable employment for a certain class of the insane, the washing is done by hand. Water is obtained from a well, from which it is pumped into a reservoir that has a capacity of 800,000 gallons. The asylum is also connected with the Sheffield water system.

Employed in various ways were 240 men and 360 women. In the tailor's shop were nine patients, including one epileptic, under a paid worker who had been for a number of years in the place he then occupied. He said that he had never experienced any difficulty in controlling the patients.

Three sewing-machines were here in use. In the shoemaker's shop, which is fitted up with the usual appliances, seven men were at work. In the carpenter's shop were two, and in the painter's shop were four patients—all working industriously. Ten men were employed in the upholstery department. Dr. Mitchell, the resident Superintendent, said that within his knowledge no patient in the mechanical department had ever inflicted injury upon himself or others.

About one hundred of the men in favorable weather work out-of-doors upon the farm and grounds, which contain in all 193 acres. About twenty-five acres are irrigated by sewage distributed from several large tanks. Cottages are provided for the bailiff and cowman. With the latter are lodged ten patients who take their breakfast and tea with him, but dine with the patients in the general dining-room. A portion of the asylum is two stories high, the upper one being used for dormitories and the lower one for day-rooms.

Separate hospital accommodation is provided for men and women. *Post-mortem* examinations are made of the bodies of about three fourths of the patients who die, permission having first been obtained from friends of the deceased.

The average of attendants to patients throughout the asylum is as one to ten. The female attendants wore black dresses, with white caps and aprons. A uniform dress gave a pauper appearance to the women patients, who were, for the most part, attired in blue print gowns and check aprons. The male patients were dressed in accordance with their several employments.

From the terraces of the airing-courts an extended view is obtained beyond the sunken walls. The number restricted to a limited area for outdoor exercise was large. Comfortable scarlet hoods are worn by female patients when they are out-of-doors in cold weather.

The capacious dining-hall, with its three large bay-windows, is used for purposes of entertainment. It is provided with a stage, gallery, and accommodation for a band which is made up of sixteen attendants. On the stage was a good piano. This hall also serves as a reading-room, having chairs, side-tables, and books. A library, to which the patients have access, is under the direction of the chaplain. The chapel, where services are regularly held, accommodates about four hundred patients.

A few unteachable idiots are among the inmates of this institution. Here, as in many other places, the Superintendent complained that the large number of chronic cases hindered his asylum from properly exercising its curative functions.

Since my visit to this asylum it has been enlarged, so that at the date of January 1, 1887, it contained 1,543 patients. Forty-seven of these were private patients, for whom a weekly charge of from 14$s.$ to 20$s.$ was made. The average weekly cost of maintenance during the year ending January 1, 1887, was 8$s.$ $\frac{1}{2}d.$ per capita.

LANCASTER COUNTY ASYLUM—PRESTWICH.

This is one of four pauper asylums that meet the wants of a county having the largest population of any in Great Britain, including, as it does, large centres of life and industry like Liverpool and Manchester. According to the last census, Lancaster contained a population of 3,485,611. On the 1st of January, 1887, it was estimated to have 83,321 paupers, of whom 10.23 per cent were lunatics. The asylum at Prestwich is about fifteen minutes' ride by rail in a northeasterly direction from the great cotton manufacturing city of Manchester. The buildings, which are of brick with slate-roofs, are for the most part three stories high. They

are surrounded by ornamental grounds. Just inside the entrance gate is a neatly constructed porter's lodge, which is a distinctive feature of English asylums. Indeed, the rules respecting the porter and his duties are, in many places, as precisely and clearly laid down as are those for the guidance of the medical staff. On approaching the asylum buildings, one is pleased to see that the windows are free from bars. The patients, dressed in drab moleskin, the ordinary attire of workingmen in this part of the country, were at work about the grounds with wheelbarrows, rollers, and garden tools.

At the time of my visit, the resident Superintendent, Dr. Ley, was assisted by four medical officers. There were 529 male and 674 female patients ; and the number of attendants to patients was one to nine. An annexe intended to accommodate about 800 patients was in process of construction. This has since been completed and occupied, and the medical staff of the asylum and the number of attendants have been increased. The inmates, January 1, 1887, numbered 998 men and 1,161 women, making a total of 2,159 patients. A very few of these are private patients, who are received at a charge of from 15*s.* to 21*s.* per week.

In an inspection of the interior of this institution one is gratified from the outset. The various wards are simply but neatly decorated and appropriately furnished. In the infirmary wards for female patients were seen numerous busts and pictures, lambrequins and light draped curtains at the windows, exotic and other plants about the rooms, and tidies upon the chairs.

The convalescent wards on the women's side are fitted up similar to the infirmary wards. The comfortable furniture and numerous articles for use and ornament recall the home-like apartments of Brookwood.

In the ward for very troublesome or refractory female patients were curtains, lambrequins, upholstered furniture, statuary, pictures, busts, and decorations similar to those in the other wards. Notwithstanding the class of patients, the open fires here are only partially screened. The windows are large and without bars. In connection with this ward is a spacious open court, in which is a small conservatory.

The furnishing and general arrangements of the apartments for both sexes are very nearly alike, except that the interior decorations on the women's side are more elaborate than those on the men's side. A large proportion of the beds have woven-wire bottoms, horse-hair mattresses, scarlet counterpanes, strong white linen sheets, and two pillows to each. For the single rooms there are double doors, the inner one having an aperture in the panel to facilitate supervision. Open fires throughout, even in the wards for suicidal patients, are only partially protected by a guard. This method of heating is supplemented by a system of hot-water pipes. The older windows have iron sashes with panes 5 x 14 inches; but wooden sashes with panes 8 x 12 inches are gradually superseding them. The hand-rails in the stairways are sunk in the wall, thus fulfilling the ordinary purpose of support without projecting.

In a new dormitory, constructed on modern and improved principles, the bedsteads are of iron and have woven-wire bottoms, upon which are laid block-mattresses made of silk-waste. Beside each bed was a thick red rug. In the sleeping apartments for the epileptics, a low form of bedstead is used. Near each bed was a stand fitted with shelves. A variety of simple yet tasteful articles ornamented the rooms.

There are one or two padded rooms, which, it was said, are rarely used. Seclusion is sometimes resorted to. In the event of the failure of moral suasion, a refractory patient is

placed in charge of a staff of attendants large enough to overcome any resistance.

The dining-halls of this institution are spacious and are elaborately furnished. That for the women accommodates upwards of three hundred. At one end is a stage, indicating that the place is also available for entertainments. At the opposite extremity is a gallery for the orchestra. The dining-tables are regularly arranged in the body and along the sides of the hall. Each table has its spotless linen cover, and chairs for ten. The crockery used is of the ordinary table pattern. Through the high arched roof of glass there is admitted an abundance of light. The gas-jets have bright colored shades. Arranged about the hall are numerous life-sized busts and figures and a great variety of luxuriant plants and flowers, among which are canary-birds in cages. A general survey of the apartment gratifies the eye by a diversity of pleasing, yet not costly, objects; while the orderly arrangement and the scrupulous cleanliness manifested not only here, but throughout the institution, give evidence of excellent administration. It is worthy of special note that the ornamentation of this beautiful hall, as well as of other parts of the establishment, was done by the patients who worked with and were directed by a few skilled artisans. The dining-hall for men is somewhat smaller than that for women, but is scarcely less elaborate in its furnishing and decorations, both apartments being well adapted to interest and gratify the patients. In the infirmary dining-room, the walls are papered. Small tables and chairs are used here, as in the hall just described. As would be expected, the diet in this department is of a better quality. Beer brewed on the premises is allowed to all the inmates. Non-workers receive half a pint per day, while workers have double that quantity.

This institution, like most other English asylums, labors under the great disadvantage of having a small estate. Additions have been made within a few years at a cost of upwards of $1,000 per acre; but even now there are only 150 acres connected with the asylum, fifty-nine of which are covered by buildings, or used as airing-courts or grounds for exercise. Twelve acres of the land are leased at a high rental. At the time of my visit, about 250 men worked either on the land, in the workshops, or in the offices, in addition to 113 ward cleaners. Twenty-nine insane men reside at the farm-house and six in a detached cottage. Of the women, upwards of 200 were engaged in attending to the wards; 38 in the laundry; 52 in the kitchen and offices; and 225 at various kinds of needlework. All the tailoring, dressmaking, shoemaking, and repairing for the establishment is done on the premises.

In suitable weather parties of male and female patients in charge of attendants take walks beyond the asylum grounds. Besides the weekly dancing parties, there are, in winter, theatrical and other entertainments.

A chapel in which Church-of-England service is held every Sunday is conveniently situated on the grounds. Mass is celebrated weekly within the institution for Roman Catholics, and a Rabbi conducts religious services at stated times for the Jewish patients.

Of those admitted to this asylum Dr. Ley has taken much pains to tabulate, from time to time, the causes of insanity, having gathered his information from examinations of histories of the patients, and from inquiry of and correspondence with the friends and relatives. The facts which he has thus collected and the conclusions he has drawn therefrom are very suggestive. He says of his investigations made during the year 1885:

"In about 27 per cent no cause could be assigned, as nothing reliable could be ascertained in regard to the antecedents of those patients. Classifying the assigned exciting causes as mental and physical, the mental causes constitute 23 per cent, and the physical about 55 per cent. Prominent among the former are worry, anxiety, and domestic and pecuniary troubles. Of the physical, intemperance in drink is preëminent, reaching about 25 per cent of all causes, and next to this comes ill-health, epilepsy, and disorders peculiar to the female sex. It will, however, be observed that a large proportion of cases in which special exciting causes are assigned, have in addition strong hereditary predisposition to the disease. At least 28 per cent inherited a weakened nervous organization, too feeble to resist ordinary exposure, and prone to become deranged by various disturbing influences, social, religious, or commercial. The troubles and worries of life are rarely powerful enough of themselves to upset a healthy brain, and drink and other kindred excesses are far more likely to derange the bodily health of the individual and carry him to the general hospital than bring him to the asylum. The experience of this and of other asylums points to the fact that the special causes usually enumerated as factors in the production of insanity are, in the majority of instances, secondary to the great fundamental cause—constitutional predisposition."

The Lunacy Commissioners in one of their reports have paid a just tribute of praise to Dr. Ley for his admirable management of this institution. As has been pointed out by them, it would be difficult to find any asylum wards which equal those of Prestwich in comfort or are more tastefully decorated. The objects of art used for ornamentation come within the reach of patients known to be wantonly destructive, yet their beauty seems to be a sufficient safeguard, and very rarely is any mischief done to them. The Commissioners say:

"It is satisfactory to add that all these great improvements are executed solely by patients' labor, superintended by skilled artisan attendants, whilst all the furniture, fern cases, etc., are obtained by the same means, as well as the busts and pottery which adorn the walls. We cannot too highly express our gratification at the appearance of the wards, and we must add that we are astonished to find that all this is done at a per-capita cost of no more than 8s. 2d. per week. We doubt if any other asylum in the country has any thing approaching to the comfort and even luxury provided here at such a rate."

LANCASTER COUNTY ASYLUM—WHITTINGHAM.

The asylum at Whittingham, opened in 1873, was planned to accommodate 1,100 patients. It is situated in the country about six miles north of Preston. The buildings are of red brick, and are one, two, and three stories in height. They cover a considerable extent of ground, and are connected by one-story corridors. The original estate contained only 155 acres of inferior land, which was bought for about $276 per acre. Small additions have since been made by purchase, and some land has been leased for the use of the asylum. The building accommodation proving insufficient to meet the growing demands upon the institution, it has been found necessary to erect an annexe for the separate treatment of chronic cases. At the time of my visit the inmates numbered 1,260, of whom 550 were men and 710 were women. The annexe has since been completed, and at the date of January 1, 1887, there were 1,679 patients.

The dormitories are on the associate plan, some of them containing as many as forty beds. For greater privacy, the beds, which occupy the central space in the room, are separated from one another by partitions four-and-a-half feet high, so arranged as to form compartments or stalls opening towards the windows. Wooden bedsteads are used. A bright rug was spread in front of each bed. The bare appearance of the walls was somewhat relieved by painting, and decorations about the windows. The lower sections of the windows had screens on the inside. Thermometers were hung in each ward.

The day-rooms were plainly furnished. The walls were plastered and painted, and brightened by colored stencilling. They were adorned with illuminated mottoes, pictures, and plaster casts. Near the windows were flowers in hanging-baskets. The window-sashes were hung with weights and cords. On the floors of the long connecting corridors were laid strips of carpet. Though not generally approved in English asylums, metal ware instead of crockery was used in the lavatories. The infirmary was comfortably and appropriately furnished. Low beds are provided for epileptics. They have a special night watch.

There are two padded rooms—one on the men's and one on the women's side. In reply to an inquiry respecting restraint, it was said: "We have no restraint. Occasionally we put on gloves, but not very often. We use no cribs, nor have we ever used them." Dresses buttoned behind are worn in certain cases. Sometimes two or three attendants hold a violent patient rather than seclude him. Wet packing is also occasionally resorted to. The patients were well clothed, and there was some variety in the style of their dresses.

There is a very large general dining-room here, which serves also as an amusement hall. The food appeared to be sufficient in quantity, and fair in quality. Crockery and blunt-pointed steel knives were used at table in the ordinary dining-room, but not in the refractory wards.

The airing-courts, to which a large number of patients were restricted, are fairly extensive. Selected parties of patients in charge of attendants are allowed to go into the country for exercise in suitable weather. There are long walks laid out within the boundaries of the asylum estate, besides ample tennis and cricket grounds. Athletic sports and dances on the green in summer, and theatrical entertainments and concerts during the winter months are included in the routine of exercises. Dr. Wallis strongly advocates the providing of amusements for the insane. He says:

"They lighten the tedium of the winter months, and make a pleasant break in the monotony of asylum life (a thing most desirable in itself for both patients and officials), and are only second, in my opinion, to good food, rest, and occupation, in furthering the recovery of the curable cases, while they reconcile, to some degree, many hopeless and weary patients to their lot, and cause them for a time to forget their troubles. The prospect of attending these entertainments acts, too, as a most potent stimulus to good behavior in many of our turbulent and troublesome patients; nothing is more likely to bring out their powers of self-control than the dread of being prohibited from attending the next theatrical performance or tea-party. The reconstruction and enlargement of our stage, recently carried out, has enabled us to undertake successfully more ambitious plays than heretofore."

On an average, about 53 per cent of the patients here are variously employed. In order to give occupation to a certain class of excitable patients, much of the washing is done by hand, a large number of ordinary wash-tubs being used for this purpose. A considerable force of men in favorable weather work upon the land, about one hundred and fifty acres of which are devoted to farm and garden purposes. The crops raised are oats, hay, and a variety of vegetables, including turnips and potatoes. The average weekly cost of maintenance during the year ending January 1, 1887, was 7s. 10½d. per capita.

The method of heating was in a transition state. The authorities were trying the experiment of substituting hot-water pipes for open fires. The effect was quite marked. Under the new arrangement an expression of gloom was depicted on the faces of the patients as they stood in their comparatively cheerless apartments gazing forlornly upon the black registers which had taken the place of the bright open fires. It was manifest that the depression was felt even by the attendants. On the other hand, the rooms where the open fires were still in use, presented a pleasing contrast, the glow of the warm fire seeming to impart cheerfulness to both patients and attendants. This was the only place in England where I saw any attempt to economize by the substitution of a hot-water system of heating for open fires, and the experiment was, to say the least, discouraging.

About five hundred patients attend divine service on Sunday, separate places being provided for Church-of-England and Roman Catholic worship.

Here, as in some other of the huge English asylums, the disadvantage of bringing together large numbers of the insane was strikingly apparent.

BIRMINGHAM BOROUGH ASYLUMS.

The Winson Green Asylum stands on a somewhat elevated site in the suburbs of Birmingham. The buildings are of brick and are mostly three stories high. The inclosed grounds, containing some forty acres, are well kept. The institution is designed chiefly for pauper patients, and is managed by a Committee of Visitors appointed by the Council[1] of Birmingham, of which committee the Mayor is a member. As in many other English asylums, an excessive number of chronic patients have been found to interfere with the curative functions of this asylum, and, at the

[1] By the Act of 1853 and that of 1855, the Councils of boroughs having a Quarter Sessions, recorder, and clerk of the peace, by giving notice to one of her Majesty's principal Secretaries of State of their intention so to do, thereby assume all the duties and powers conferred by statute upon the justices of boroughs in respect to the erection and management of borough asylums.

time of my visit, a new one, designed to accommodate some eight hundred chronic and imbecile cases, was in process of erection. This is situated at Rubery Hill, eight miles out of Birmingham. Owing to limited accommodations, it had been found necessary to contract with other institutions for the reception of patients. The provision made here for the pauper insane is not inferior to that for indigent patients of the non-pauper class.

The several wards were well furnished. In the sleeping apartments, there was noticed in front of each wooden bedstead a strip of carpet. The furnishing of each bed comprised two mattresses and a fair supply of bed-clothes. The interior brick walls, although for the most part unplastered, are pleasingly tinted and stencilled. In the sitting-rooms the walls were papered and adorned with pictures. The windows were provided with lambrequins, and the floors covered with linoleum. In some of the apartments were carpets and cushioned lounges. Epileptic patients are under special supervision. There are two padded rooms in the institution—one for each sex. In the refractory wards, the floors are of tile. The windows have iron sashes, with panes measuring 5 x 8 inches. It was said that mechanical restraint was resorted to only for surgical reasons or for patients inclined to commit suicide, and that occasionally it was found necessary to have recourse to seclusion.

Those able to work are provided with employment to the extent of their ability. Six patients were engaged at shoemaking. One of the industries carried on here is that of match-box making. A considerable number were engaged in keeping the grounds in order.

The attendants were uniformly dressed, the women wearing dark gowns, and white caps trimmed with green ribbon. The dress of the patients seemed fairly comfortable, though some of the men appeared too thinly clad considering the severity of the weather. On the men's side, are three confined airing-courts used respectively by infirmary patients, idiotic children, and the refractory. There is also one general airing-court.

The Medical Superintendent, whose experience extended over a period of nearly forty years, spoke of the great changes that had taken place in the treatment of the insane within the memory of living men. He mentioned having seen as many as fourteen men in a single room chained to a wall. Even the fire-irons were in his earlier days chained to prevent personal injury. Then, the bare suggestion of permitting an insane man to use a knife and fork at meals, would have filled the mind of an asylum superintendent with consternation. Spoons even were forbidden, but after a time were sanctioned; and opinion gradually changed until at length the use of the knife and fork was allowed. "I had long desired," he said, "to banish from my practice slavish adherence to mechanical restraint in every variety of form which then obtained; but I was strongly opposed by the managing Committee. Indeed, my repeated appeals in this direction got me into rather bad favor with them. It so happened, that, during one of their meetings, a patient was reported to be in a condition of great excitement. He had fortified himself so strongly in his room that no attendant dared enter. I was jocosely asked by the Chairman of the Committee to put this patient under one of my non-restraint experiments. The challenge was accepted. I went to the ward, found the patient strongly barricaded, and threatening death to all comers. In spite of the entreaties of friends and attendants, I made my way alone into the room beside him. Knowing his great fondness for tobacco, and pretending not to notice his disturbed condition, I proposed a smoke, at the same time offering him some tobacco. The ruse was entirely successful. The patient instantly calmed, and permitted himself to be disarmed and conveyed to another ward. From that date, the managers of the institution ceased to scoff at the practicability of restraint by other means than that of irons."

In this old asylum, unrelieved by many modern improvements, there appeared to be little worthy of imitation. The yards, or airing-courts, were small and prison-like; the amount of land and other requirements insufficient for the wants of a large and populous district like that of Birmingham; and the attendance quite inadequate to give effect to the advanced theories of the Superintendent.

Since my visit, drill, calisthenic exercises, and singing classes have been introduced here. Their introduction was the result of the favorable report of a special committee who visited the Richmond Asylum, Dublin, where these exercises form a part of the educational system introduced by the late Dr. Lalor. Other changes have taken place and improvements have been made by the new Superintendent, which have raised this institution to a higher standard. The asylum at Rubery Hill has been opened, and, January 1, 1887, it contained 574 patients. This has relieved the overcrowding at Winson Green, which, on January 1, 1887, had 575 patients. The per-capita average weekly cost of maintenance during the year ending January 1, 1887, at Winson Green was 8s. 11¼d.; at Rubery Hill, 8s. 8d.

THE FRIENDS' RETREAT—YORK.

Projected in the spring of 1792 and opened in 1796, the York Retreat is memorable as the place where the non-restraint principle was first adopted in Great Britain—where William Tuke, on behalf of the Society of Friends, courageously renounced, as did Pinel in Paris, the use of chains and manacles in the treatment of the insane.

This asylum is about a mile from the city of York. An ivy-covered wall incloses the extensive and somewhat elevated grounds, which are artistically laid out and carefully kept. The Retreat, a private institution, or "registered hospital," is still directed by those who are, for the most part, members of the Society of Friends. It has a yearly income of about $75,000. The buildings are old-fashioned, but wear an aspect of unpretentious comfort. In the oldest part, some of the arrangements scarcely meet all modern requirements. "The Lodge," a building recently erected for men, is constructed on modern principles, and furnished in a style approaching luxury. Eight women occupy a sumptuously furnished detached residence called "Bellevue House." There is also a neat villa residence for the reception of a few female patients.

In the wards of the male and female divisions the beds and bedding appeared to be clean and comfortable, and the apartments were warm and well-ventilated. The corridors, with tinted walls and bright embellishments, were quite pleasing. The windows in the oldest portion have iron sashes, with panes $6\frac{1}{4} \times 7\frac{1}{2}$ inches; in the newer portions, the panes are larger, and the sashes are adjusted by weights and cords. The patients have their meals in their several wards.

A large sitting-room, in which were noticed a fernery and a variety of other attractions, is used on Sunday evenings for religious services, at which are present those of widely different beliefs. Patients to whom it is proper to extend the privilege are permitted to attend public worship in the city, in the churches of the various denominations to which they belong. A new association room suitably furnished accommodates one hundred and fifty persons, and here lectures and entertainments are periodically given.

The female patients assemble once a week for reading, music, games, and social intercourse. Permission is occasionally granted to some of the patients to visit places of amusement in York.

At the time of my visit, the records showed 153 patients —64 men and 89 women. The accommodation afforded is dependent more upon mental condition than upon the scale of payment. While some are provided for gratuitously, the prices paid for maintenance of others range from trivial sums to as high as £400 per annum. The average cost of support is about 32s. per week.

Two padded rooms—one for each sex—are provided in case of need. These are so fitted up as to obviate the possibility of injury. Locked dresses are not used, although strong garments are sometimes put on when patients are sent to their rooms during paroxysms. Order prevailed throughout the asylum, and the surroundings were quiet.

A feature of the management is the "autumnal trip" made every year to the sea-side by patients in charge of the Superintendent or some of his staff. Instead of, as heretofore on such occasions, procuring accommodations in expensive lodgings, a house has recently been leased in Scarborough, where those likely to be benefited by a change may enjoy the salubrious air of the Yorkshire coast.

A full inspection of the Retreat, with its detached cottages, its artfully screened walls, its beautiful recreation grounds, and its wealth of flowers within and without, suffices to convince the visitor that this small and select institution, which well-nigh a hundred years ago proved such a powerful factor in educating and elevating public opinion in England, retains to this day much of that progressive spirit and humanity of purpose upon which its world-wide reputation rests.

CHAPTER III.

SCOTLAND.

THE present admirable Scotch lunacy system dates from August 25, 1857. Looking backward from that time, we see little in the history of the treatment of the insane in Scotland that is worthy of commendation, if we except what was accomplished in their behalf through private benevolence. The peculiar superstitions of the people at one time contributed to the record of cruelties perpetrated upon this unfortunate class.

Allusion has already been made to the fact that the insane were sometimes put to death under the belief that they were witches. The last instance of this kind in Scotland is recorded by Sir Walter Scott as having taken place in 1722, and is thus described: " A sheriff-depute of Sutherland, Captain David Ross, of Littledean, took it upon him, in flagrant violation of the then established rules of jurisdiction, to pronounce the last sentence of death for witchcraft which was ever passed in Scotland. The victim was an insane old woman, belonging to the parish of Loth, who had so little idea of her situation as to rejoice at the sight of the fire which was destined to consume her. She had a daughter lame of both hands and feet, a circumstance attributed to the witches having been used to transform her into a pony, and get her shod by the devil." It does not appear that any punishment was ever inflicted for this cruel abuse of law on the person of a creature so helpless. Scott also relates that the son of the lame daughter, he himself distinguished by the same misfortune, was in later years a beneficiary of the Marchioness of Stafford.

As an illustration of the inhuman treatment of the insane at a later period, one method of incarceration is here given, as described by Dr. Robert Gardiner Hill: "At Inverness, between the second and third arches of the old bridge, built in 1685, there is a dismal vault, used first as a jail and afterwards as a mad-house. This appalling place of durance, where the inmates were between the constant hoarse sound of the stream beneath, and the occasional trampling of feet and rattling of wheels overhead, existed as late as 1815, and is said not to have been abandoned till its last miserable inmate, a maniac, had been devoured by rats."

During the latter part of the eighteenth and the early part of the nineteenth century the state of the insane poor was one of wretchedness and neglect. In 1818 there were in all Scotland only about two hundred and fifty insane persons provided for in public asylums and about one hundred and fifty in private asylums. Twenty years later the condition of this class was not much improved. There was a lamentable insufficiency of asylum accommodation. Large numbers of the insane were shut up in private dwellings, without official supervision; others were restricted to poorhouses with no special provision for their care; some were confined in jails; and a few, including even the violent and dangerous, were living alone or were permitted to roam about the country.

While there existed a reprehensible reluctance on the part of the State to discharge its obligations to the insane, it is gratifying to note some noble manifestations of private generosity in their behalf. The opening of the Royal Edinburgh Asylum in 1813 was the result of individual effort.

THE BURNING OF THE LAST PERSON SENTENCED TO DEATH FOR WITCHCRAFT IN SCOTLAND, 1722.

Here, from the outset, an intelligent and humane system of treatment was pursued and an opportunity offered for the study of mental diseases. Murray's Royal Asylum at Perth was erected out of funds bestowed by the benefactor whose name it bears. The Crichton Institute at Dumfries, opened in 1839, was endowed with upwards of $500,000 left by Dr. Crichton for charitable purposes. Previous to 1855 seven chartered asylums, called " Royal," had been founded by private charity, but no government aid had been granted, except £2,000 to the Royal Edinburgh Asylum at Morningside. These institutions had expended on lands, buildings, and furniture £352,632, in making provision for both pauper and private patients. But these generous efforts were wholly inadequate to meet the wants of the insane.

The first parliamentary legislation to regulate the confinement and treatment of lunatics in Scotland was the Act 55 George III. c. 69, passed in 1815. This was followed by the enactment of statutes for the same purpose in 1829 and 1841. These several acts, which may be said to have formed the Scottish code of lunacy, were so ambiguously framed and so contradictory in many of their provisions that there was great difficulty in administering them. It was not until 1855 that any effective movement was set on foot to establish a humane and comprehensive lunacy system. And here it may be proper to mention that the people of Scotland are indebted to the philanthropic labors of Miss Dix for originating the great reform begun at that time. This estimable lady, in relating to the writer some of her experiences while endeavoring to improve the condition of the insane in foreign lands, told him of her efforts in Scotland. It appeared from her account that she was greatly shocked at the condition in which she found the insane in that country, and was much disturbed because of the difficulties she encountered in gaining desired information. Finally she was refused admission to an asylum that she particularly desired to visit at night, in order to learn of peculiar abuses that she believed were hidden from the public. She applied to a high official in Edinburgh for the privilege denied her by those in charge of the institution. Being again refused, she proceeded at once to London and laid before the Home Secretary the startling facts of which she had become possessed respecting the sad condition of the insane in Scotland. Notwithstanding the sturdy opposition made by her conservative opponents, the result was the appointment by Parliament, on the 3d of April, 1855, of the famous Royal Commission, "to inquire into the condition of Lunatic Asylums in Scotland, and the existing state of the law in that country in reference to Lunatics and Lunatic Asylums."

The feeling aroused led to very earnest work on the part of the Commissioners, who were engaged two years in making investigations and preparing their report. This brought to light the imperfections of existing laws, exposed the cruelties to which the insane were subjected, and opened the way for the excellent lunacy system Scotland now enjoys.

Concerning the chartered asylums, the Commission reported that they were "in many respects in a highly satisfactory state," that the management was disinterested, and that "the treatment of the patients was liberal and judicious," and, "on the whole, deserving of commendation." In these institutions the monotony of asylum life was broken by frequent excursions, occasional picnics, concerts, lectures, evening parties, and dances, and in a few, by trips to the sea-side in summer. There was some adverse criticism made, however, on these asylums. Some of them were overcrowded and were imperfectly warmed and ventilated. Refractory patients were secluded in dark cells with stone floors. In the Montrose asylum dirty and destructive patients were permitted to remain entirely naked while in seclusion, loose straw being cast on the floor for bedding. Complaint was made of the general insufficiency of land attached to asylums, the absence of outdoor as well as varied indoor employment, and a lack of many things now considered essential in asylums for the insane.

In many of the private asylums or licensed houses the Commission found the condition of the insane to be most deplorable, the principal aim of the proprietors being to secure increased profits. The patients were crowded into small apartments, using the same rooms by day and by night, with no proper separation of the sexes, and, to save candle-light, kept in bed the greater part of the twenty-four hours during the winter months. The bedding was usually of a coarse and cheap description, insufficient in quantity, and not renewed when filthy. It was stated that in one asylum women patients were stripped naked at night, and two, and sometimes three were made to sleep in the same bed-frame on loose straw, in a state of perfect nudity. The proprietor of another establishment, testifying before the Commission, admitted that the floors at night were soaked with wet; that half the dirty patients slept naked; that seven or more would sleep together in a state of nudity; and yet he considered this proper treatment for them. Comment was made on the undue use of restraint, the lack of attendants and opportunities for proper exercise, and the insufficiency of food and clothing.

The condition of the insane in the poorhouses was, as might be imagined, unsatisfactory in the extreme, and was the subject of severe criticism.

Library of
Robert Jaudon Ball

The report, while stating that the details furnished gave but an imperfect representation of the true condition of the insane outside of institutions, presented an appalling picture of the amount of misery suffered by this class throughout Scotland.

It is true the law provided a system of inspection; but the officers entrusted with this duty were strangely derelict, and the Commission found it necessary to report that most radical measures, amounting to an entirely new system, were requisite to meet the urgent needs of the time.

Public attention was directed to the desirability of building asylums of moderate size; to the great benefits of fresh air, healthful exercise, and profitable occupation; and also to the disadvantages attending the then too frequent prison-like construction of institutions for the insane. The Commission recommended that asylums should be more home-like in appearance; that strong wire and iron enclosures should be abolished; and that more attention should be paid to the principles of heating and ventilation.

The outcome of this report was the Act 20 and 21 Vict. c. 71, passed August 25, 1857, entitled "An Act for the Regulation of the Care and Treatment of Lunatics, and for the Provision, Maintenance, and Regulation of Lunatic Asylums in Scotland." In accordance with this memorable Act, a General Board of Commissioners in Lunacy for Scotland was appointed, and the country was divided into asylum districts. It was required that asylums for the insane poor should be erected where necessary, and provision was made for their supervision and control.

At present the lunatics[1] of Scotland are provided for in Royal, district, and parochial asylums, in the lunatic wards

---

[1] As in England, the statutory meaning of the term lunatic includes every person of unsound mind and every person being an idiot.

of poorhouses, in private asylums, in private dwellings, in training schools for imbecile and idiotic children, and in the lunatic department of the State Prison at Perth.

The seven Royal asylums are, strictly speaking, charitable institutions, managed by unpaid boards, who have no pecuniary interest in their prosperity. All, except Murray's Royal Asylum at Perth, admit pauper as well as private patients, the board of the former being paid by parochial authorities, and that of the latter by friends, from which sources these institutions mainly derive their support.

The district asylums, of which there are twelve, are public institutions, and were built from assessments. They are managed by district boards, and receive only pauper patients, unless there is an excess of accommodation. Where there is an excess, it may be used for patients of small means, who pay but little more than the rate paid by the parishes of the district for their pauper patients.

The parochial asylums, numbering six, have been erected out of taxes levied upon the parishes to which they belong. They are managed by parochial boards, and receive only pauper patients.

There are sixteen lunatic wards of poorhouses having restricted licenses, in which are received only harmless and chronic cases. These wards are, in some instances, separate from the poorhouses, and may be regarded as detached asylums.

There are six small private asylums, kept either by medical men or laymen, which are licensed by the Lunacy Board to receive a limited number of patients. The private asylums receive only private patients, and are required to pay £15 each, annually, for a license.

Besides the above-named receptacles for lunatics, both private and pauper patients are received into private dwellings under the sanction of the Commissioners in Lunacy. Not more than four patients can be admitted into any private house, and a license is required if more than one is received.

The largest asylum in Scotland is designed to accommodate not more than eight hundred patients, the others being built for a less number, in accordance with a prevailing opinion in favor of small institutions.

In construction, as well as in management, the public asylums have, during the past few years, lost much of their institutional character, and have correspondingly gained in public estimation. Walled courts, high fences, locked doors, physical or mechanical restraint, and the use of stimulants and narcotics are, to a great extent, discarded. Among the reforms are closer and more intelligent supervision, the substitution, to a large degree, of moral for physical restraint, and a great increase of industrial occupations. Considerable tracts of land have, during late years, been added to several of the more important institutions. Farm labor is found to be highly beneficial in the treatment of curable as well as incurable patients, besides proving an economical source of supply to the asylums.

The headquarters of the General Board of Commissioners in Lunacy for Scotland, which is the chief authority in lunacy matters, are in Edinburgh. The Board consists of five commissioners, who are assisted by a secretary and a subordinate force of clerks. Three of the commissioners, including the chairman, are unpaid. The paid commissioners are particularly charged with the work of visitation and inspection. They have always been medical men, though it is not specified by the statute that they shall be. Two of the unpaid commissioners have always belonged to the legal profession, although this is not required by law; and at present a baronet, who is also unpaid, represents the laity in the capacity of chairman. There are two Deputy Commissioners whose special duty it is to visit the insane in private dwellings.

By the action of the legislature large powers are given to the Lunacy Board. These extend, in a greater or less degree, over the care and treatment of all lunatics in public and private asylums, in the lunatic wards of poorhouses, and all pauper patients in private dwellings. The supervision of the Board also extends to every non-pauper insane person who is kept in a private house for profit and suffers from mental disorder of a confirmed character; or who, whether kept for profit or not, has been insane for more than a year and is subject to compulsory confinement to the house, to restraint or coercion of any kind, or to harsh and cruel treatment. The supervision of the Board also extends to those patients who are possessed of property which has been placed in charge of a judicial factor by a court of law.

The Lunacy Commissioners are authorized to grant, renew, or suspend the licenses of private asylums, parochial asylums, and the lunatic wards of poorhouses. They are empowered to institute inquiries, to summon witnesses with the concurrence of the Lord Advocate, to examine them on oath, and to direct payment of their reasonable expenses. They, or at least two of them, inspect not less than twice a year all chartered, district, parochial, and private asylums and all the lunatic wards of poorhouses, as also the institutions for idiots and the department for insane criminals in the prison at Perth. They are authorized to employ such medical assistance as may be necessary, and are required to make annually a report of their work and proceedings to the Secretary for Scotland, for transmission to Parliament. The Lord Advocate, who is the chief law officer for Scotland, has a right to inspect all the books and proceedings of the Board of Commissioners.

The procedure under the Scotch law in consigning patients to asylum care is the same whether the person to be admitted is possessed of property or is a pauper, and whether he is to be placed in a public or in a private asylum. The statute requires that the officials authorized to place persons in asylums must be those to whom are entrusted the power of taking away personal liberty for other reasons than that of insanity. Therefore it becomes necessary to obtain the order of the sheriff [1] before a private or a pauper patient can be permanently received into any asylum. To secure this, some person must petition the sheriff to grant such order, and he must state the relation in which he stands to the patient. The petition must be accompanied by a written statement of particulars signed by the applicant. It must also be accompanied by two certificates granted by registered physicians, setting forth that they have separately examined the patient, and found him to be a lunatic and a fit person to be placed in an asylum. The sheriff's order must be acted on within fourteen days after it is granted, unless it is issued in the Orkney or Shetland Islands, in which event twenty-one days are allowed. The petition for the reception of a pauper patient into an asylum is usually presented to the sheriff by the Inspector of Poor.

Within seven days after the sheriff issues an order for the reception of a lunatic into any asylum or house, the sheriff's clerk must send notice thereof to the Lunacy Board, stating by whom the application was made, to whom the order applied, giving the names of the medical men who signed the certificates, the name of the sheriff who granted the order,

---

[1] Under the statute the term sheriff includes sheriff-substitute.

and the name of the asylum or house to which the patient is to be admitted.

Under pressing circumstances, when there is not time to obtain the sheriff's order, a patient may be received into an asylum on a certificate granted by one registered medical practitioner, declaring that the person named in the certificate is of unsound mind, and a proper person to be placed in an asylum, and that the case is one of emergency. This certificate may be granted by the medical officer of the asylum into which the patient is to be received. Within three days, however, the sheriff's order must be obtained; otherwise the detention of the patient becomes illegal, and he must be discharged.

When it comes to the knowledge of an Inspector of Poor that there is a pauper lunatic within his parish who has not been "intimated" to the Lunacy Board, he must within seven days inform the Commissioners. The Inspector of Poor must also report the case to the Chairman of the Parochial Board.

The insane are not admitted into the lunatic wards of poorhouses unless these departments have been licensed by the Commissioners in Lunacy, and are governed by rules and regulations approved by them. Nearly all of the inmates of the lunatic wards have been former inmates of asylums, whence they have been transferred by an order of the Lunacy Board granted on the petition of the Inspector of Poor. They may, however, be admitted directly by an order of the Lunacy Board. Both in the case of transfer and of direct admission there must be a medical certificate stating that the patient is harmless and does not require asylum care. This must be given by some other than the medical officer of the poorhouse in which it is proposed to place the patient, and it is in addition to the two establishing his lunacy, which were obtained when he was first intimated to the Board. Patients may also be admitted into the lunatic wards of poorhouses on the order of a sheriff, the usual medical certificates having been obtained; but as the consent of the commissioners is likewise necessary, this mode of procedure is rarely resorted to. No patient can be permanently admitted except by the sanction of the Lunacy Board.

Insane persons arrested by the police as dangerous or offensive to decency, on petition of the public prosecutor are tried before the sheriff, and if found to be insane, are committed by him to asylum care. These constitute a very small percentage of the insane.

In the case of persons of property becoming insane, the highest legal tribunal, namely, the Court of Session, is petitioned upon medical evidence to appoint a judicial factor to take charge of the property of the patient. This having recourse to a legal tribunal whose sittings are in Edinburgh, proved expensive and troublesome in the case of very small estates. In 1880, therefore, a law was passed conferring upon sheriffs or their substitutes in the several sheriff courts in Scotland jurisdiction in all such cases where the yearly value does not exceed £100.

The commissioners have power to grant orders for access of friends to patients, and these orders must be respected on pain of a fine not exceeding £20.

In 1862 an act was passed permitting the admission to asylums of what are termed "voluntary patients"; but it proved to be so cumbersome that few availed themselves of its provisions. In 1866 this measure was repealed, and another having a like object, but with fewer restrictions, was enacted. Under the law as it now stands voluntary patients are not registered as lunatics, and no medical certificate is necessary for their reception. The patient must make the application himself by letter addressed to the Lunacy Commissioners specifying the asylum he wishes to enter, and with their sanction the superintendent is authorized to admit the applicant. After becoming an inmate, the patient is permitted to leave the asylum at any time by giving three days' notice. Some persons admitted as voluntary patients are afterwards found to be insane, and while remaining in the asylum are regularly committed by a sheriff's order. In such cases, however, a special examination is previously made by the commissioners.

Respecting the operation of this law the commissioners say in a late report:

"We have for some years been able to state that nothing has occurred to indicate any difficulty or disadvantage traceable to the presence of this class of patients in asylums; and we continue to be of opinion that it is a useful provision of the law which permits persons who desire to place themselves under care in an asylum to do so in a way which is not attended with troublesome or disagreeable forms, but which nevertheless affords sufficient guarantee against abuse. At the visits of the medical commissioners to asylums all voluntary inmates are seen, and they have then an opportunity of making statements in regard to their position, should they desire to make any. When there is reason to suppose that they in any way fail to understand the conditions of their residence, we consider it proper to explain these conditions; but we have never found that the nature of their position has been intentionally concealed from them."

The manner of releasing patients from asylum care in Scotland differs materially from that of England, and is worthy of careful consideration. The Scotch method has been succinctly explained by Dr. Mitchell, one of the Commissioners in Lunacy for Scotland, as follows:

"The very fact of restoration to sanity makes the detention of a patient illegal. No particular procedure, however, is laid down by the law for such a case. The superintendent simply discharges the patient as recovered, and gives notice of the discharge to the Board. If the patient thinks that he is still detained, though he has reached a state of sanity, he can appeal to the Board, who can order his discharge, on being satisfied by the certificate of two medical persons, whom they may think fit to consult, of his recovery or sanity. The Board cannot order the discharge of any patient of whose complete restoration or sanity they are not thus satisfied ; but it is a provision of the law that any person having procured and produced the certificate of two medical persons, approved by the sheriff, either of the recovery of any patient, or bearing that the patient may, without risk of injury to himself or the public, be set at large, may petition the sheriff to order his discharge, which order the sheriff is empowered to grant. This procedure relates both to recovered and unrecovered patients. The Board can thus only discharge recovered patients, but the sheriff can discharge both recovered and unrecovered. The commissioners and the sheriff alike require to have the condition of the patient testified to by medical men of whom they approve. The commissioners do not act on their own opinion, except in so far as it falls to them to determine whether there is, *prima facie*, a case for inquiry ; in other words, whether medical men should be sent to examine.

"Again, the sheriff's order, on which a patient is received into an asylum, is not now of unlimited duration. It does not remain in force longer than the 1st of January, first occurring after the expiry of three years from the date on which it was granted, or than the first day of January of each following year, unless the medical superintendent of the asylum in which the patient is kept shall on each of the said first days of January, or within fourteen days preceding, grant and transmit to the Board a certificate that on a careful review and consideration of the case, he is of opinion that continued detention is necessary and proper. This provision is intended to secure a careful yearly revision of the condition of patients who have been three years or more in asylums, and a formal expression of opinion regarding it, and in the opinion of the Board it is a provision which practically does good. If the certificate is not granted, the patient is *ipso facto* discharged ; the certificate especially declares that the continued detention of the patient is necessary and proper.

"Private patients can be taken out of asylums at any time by their friends; but the superintendent, if he thinks the discharge of the patient would be attended with danger, may appeal to the public prosecutor to interpose his authority and deal with the patient as a dangerous lunatic. On the other hand, a superintendent can force the friends or guardians of a patient, whether private or pauper, to remove him, if he thinks he has so far recovered that he may be liberated without risk to the public or himself. If his opinion as to the liberation of such a patient is resisted by the friends or guardians, he can appeal to the Board, who, on satisfying themselves that his opinion is correct, can order the discharge of the patient. The relations and powers of parochial authorities in regard to pauper lunatics have, by the later amendments of the Scotch Lunacy Acts, been made as similar as possible to the relations I have just described as existing between private patients and their friends. The removal of pauper lunatics by parochial authorities is not much more difficult than the removal of private patients by their friends; but when any pauper lunatic is removed as unrecovered, the Board require information from the parochial authorities as to whether parochial relief is still extended to him, and if it is, then the arrangements made for the patient's care must be submitted to the Board for their approval and sanction. He remains under the jurisdiction of the Board as a patient in a private dwelling-house, and is visited by the Deputy Commissioners from time to time. The return of such a patient to the asylum can be ordered by the Board at any time. In this way no injury can be done to pauper patients by the considerable powers which are given to parochial boards in regard to their removal from asylums; on the other hand, the advantages both to patients and to parochial boards from these powers are important."

Persons committed as dangerous lunatics are not liberated without the approval of the public prosecutor.

Every released person who may consider himself to have been unjustly confined is entitled to receive a copy of the order, certificate, etc., upon which he was committed.

The certificate of the surperintendent or medical attendant, which is necessary to detain a patient in an asylum longer than the "1st of January, first occurring after the expiry of the three years from the date on which it was granted, or than the first day of January of each following year," is in accordance with section 7 of the Act 29 and 30 Vict. c. 51, and is in the following form:

"I hereby certify, on soul and conscience, that I have, within a period not exceeding one month preceding the date of this certificate, carefully reviewed and considered the cases of the patients whose names are subjoined, and I am of opinion that their continued detention in the asylum is necessary and proper for their own welfare (or, for the public safety, as the case may be)."

Superintendents have power without the consent of the Lunacy Commissioners to liberate patients on trial for a term not exceeding twenty-eight days. This practice has proved so beneficial that it is regarded with increasing favor.

In some asylums it is customary to release patients on probation for periods, usually, of from three to six months. This can be done only by the consent in each case of the Lunacy Commissioners. During his absence from the asylum the patient is subject to the inspection of the Board, and if a pauper lunatic, his name cannot be taken off the poor-roll. If necessary he can at any time be returned to the asylum. Probationary removals were first authorized in 1862, and from that time to the close of 1886 there were 2,982 patients thus temporarily discharged, 530 of whom were returned to the asylum before the expiration of the period of probation.

Respecting this excellent provision of the Scotch statute, the Lunacy Commissioners say:

"The special use of the statutory discharge on probation is to permit of the conditional liberation of patients whose fitness for permanent discharge cannot be determined without actual trial, under the conditions of ordinary life, for longer periods than twenty-eight days. It is frequently found that patients, who appear while in the asylum to have improved so much that they are fit for being provided for in private dwellings, become unsettled when the restraints of the asylum are removed. It is not, however, justifiable to retain permanently in the asylum all patients in whose cases a possibility of such unsettlement is thought to exist. The large majority of patients discharged on probation undergo no deterioration, and many are benefited by the change. By discharging patients on probation there is an opportunity for testing their fitness for permanent discharge, and at the same time for replacing them in the asylum without the expense attending a sheriff's order, if they prove unfit for permanent discharge. We continue to be of opinion that in some establishments a more frequent application of the probationary discharge to patients whose fitness for residence in private dwellings may be uncertain, would lead to a larger number of permanent discharges than takes place at present."

A penalty not exceeding £100, or imprisonment not exceeding six months, may, without prejudice to action for damages, be inflicted upon officers or employees of any public, private, or district asylum or house, who shall wilfully maltreat, abuse, or neglect any person detained therein as a lunatic patient. The public prosecutor deals with such cases. Penalties are recoverable before the sheriff for making false statements or returns, or for refusal to give information as required by law.

Respecting medical attendance upon patients under asylum care, it is required that all asylums licensed for one hundred or more patients must have a resident physician; those licensed for more than fifty and less than one hundred need not have a resident medical officer, but they must be visited daily by a medical man. Those asylums licensed for fifty or less than fifty patients must be visited twice a week by a medical man. The Lunacy Board have power to increase the number of visits, and they may require a resident physician to be appointed to any asylum licensed for more than fifty patients. It is provided, however, that the commissioners may give permission to visit houses licensed for fewer than eleven patients less frequently than twice a week, but not less frequently than once in every two weeks.

Out of an estimated population of 3,949,393 in Scotland January 1, 1887, there were under cognizance of the Lunacy Board 11,309 lunatics. This gives the proportion of lunatics to the whole population as 1 to 349. In 1867, it was as 1 to 469. In 1858, the proportion was as 1 to 517. While the number of lunatics coming under the supervision of the Lunacy Board during the period from 1858 to the date of January 1, 1887, increased, according to the estimate of the Board, 96 per cent, the growth of the population during the same period was only 31 per cent. The commissioners state, however, that this increase does not necessarily indicate an increasing amount of mental disease, but that they consider it due, in a large measure, to a growing readiness to place persons more or less disordered in mind in lunatic establishments.

The 11,309 lunatics under the cognizance of the Scotch Lunacy Board on the 1st of January, 1887, were distributed as follows: In the seven royal asylums there were 3,184; in the twelve district asylums, 3,142; in private asylums, 128; in parochial asylums, 1,444; in the lunatic wards of poor-houses, 857; in private dwellings, 2,270; in the lunatic department of the general prison at Perth, 56; and in the training institutions for imbeciles or idiots, 228. Of the above total, 9,514 are maintained out of parochial rates, 1,739 from private sources, and 56 at the expense of the State.

During the year ending May 14, 1887, the average weekly per-capita cost of maintenance of pauper lunatics in royal, district, private, and parochial asylums, and schools for imbeciles was 10*s.* 0¾*d.*; in lunatic wards of poorhouses, 7*s.*; in private dwellings, 5*s.* 10*d.*

The government grant of four shillings per capita per week toward the maintenance of pauper lunatics, begun in 1874, has been a powerful means of raising the standard of care for the needy insane. It increased from £59,483 in 1875 to £88,258 in 1887. The terms of the grant are somewhat peculiar. It is not a uniform one of four shillings per capita. When the actual expenditure for a patient falls below eight shillings per week, only half the amount expended is allowed by the government. In Scotland this grant extends to the lunatic wards of poorhouses. An examination of these wards impresses one with their superiority to the generality of poorhouse provision, this superiority being due, perhaps, to the fact that the poorhouses are under the direction of boards of intelligent and unpaid managers; yet in no case did it appear to the writer that the standard of care was equal to that of asylums.

The following information, which further illustrates the workings of the Scotch lunacy system and its somewhat novel features, the writer obtained in an interview with Dr. Sibbald, one of the commissioners:

"Our Board is called the General Board of Commissioners in Lunacy for Scotland. Each district has its district board. Each town is governed by its municipality, which is a representative body; but outside the towns affairs are managed by Commissioners of Supply, consisting of all persons possessed of land over the value of, I think, £200 a year. They meet twice a year in a kind of small parliament, and they have the appointment of all the local boards, such as the District Board of Lunacy, also the board managing the police regulations of the district,—every thing of this kind is under their control. They appoint committees of their number to attend to local matters; and one of these committees is called the District Board of Lunacy. This board has to erect the district asylum, keep it in repair, appoint officers, and see that there is accommodation for all the lunatic poor of the district. It does not provide the maintenance for the lunatics, because each parish pays for its own ; it simply provides the building, furnishes it, and appoints officers for the district.

" Some of the districts, of which there are twenty-two, are large and others small. Sometimes a portion of a county, sometimes a county, and sometimes a congeries of counties forms a district. The members of the district boards have under their charge the building and government of the district asylums ; they decide what the supply shall be ; they present to the meetings of the Commissioners of Supply the amount required for the year ; but, on these matters, they report to us first, and we give authority to levy the assessment. They cannot levy without our approval. The boards differ in size,—the smallest numbering seven, and the largest thirteen members. They are unpaid. Practically they cannot put up a building without our approval. They have a great deal of liberty nominally ; but they are under the influence of public opinion. If they were to do any thing which the public thought wrong, then, at the meeting of the Commissioners of Supply in all probability they would be turned out and others appointed. We are required to make a full report on these asylums, and our report is printed.

" At the time the first of what are known as the Lenzie Acts was passed, certain bodies had already provided asylums. Some of them were corporations established as charitable institutions. They received a certain number who could not pay any thing, and also those who desired to be admitted and were willing to pay for their maintenance. Among these charitable foundations were the institutions at Morningside, Dumfries, Montrose, Gartnavel, and Aberdeen,—all chartered asylums. That at Aberdeen is also the district asylum for Aberdeenshire.

" There were also, at that time, some of the larger parishes which had provided asylums for their poor people. The Glasgow parishes may be cited—three large parishes, Govan being one. Govan had not made such provision, but the Barony parish of Glasgow had to a small extent. It could boast of a small asylum—not a good one, it is true ; still it was in the position of having provided accommodation for its lunatic poor. In terms of the Act of Parliament, those parishes which had so provided were to be allowed, if they made suitable arrangements, to continue to make provision, if they desired to do so. The present Lenzie Asylum is the outgrowth of that legislation. There they have the management vested in the parochial board and a committee of that board under our supervision. There is no district board to supervise them. They require only to apply to us every year for a license to keep their asylum open. If the asylum was badly conducted, it would be our duty to refuse the license. Certain other parish institutions might be named, as, for example, the asylums of Abbey, Paisley, and Greenock. These are parochial asylums like Lenzie, which provide, among them, for all the parishes of Renfrewshire, whose district board has consequently no functions. It exists on paper, but that is all. If these parishes were to fail to provide for the wants of the county in regard to the insane, the district board could be called into existence, and might make provision ; but, at present, that is not necessary. The remaining parishes of Renfrewshire have made agreement with the three which possess asylums. Each of these three, therefore, has a certain number of parishes associated with it, and forms practically, though not technically, a little district.

"In Glasgow, to which Lenzie belongs, you have a very different class of patients from what would be found in such a county as Fife, Aberdeenshire, or Inverness-shire. They are accustomed to different kinds of work, and they are generally a rougher class in the towns. The lower stratum of a town is generally lower than that of a rural district.

"With regard to asylums, each patient is admitted on application to the sheriff, who is the county judge. A statement of the case with two medical certificates is sent to our office from the asylum superintendent after the patient has been admitted. The superintendent reports upon the mental condition of the patient, so as to indicate that he considers the person to be a proper subject for asylum treatment. Therefore we have not only the assurance of the two medical men who sent the patient, in believing him to be insane, but we have the additional assurance of the superintendent. If we find the person described simply as restless and disinclined to work, we write back asking the superintendent whether he considers him insane. If the superintendent is still unable to give a more definite opinion, we say: 'Report again in a month.' If, at the end of that period, he is still unable to certify to the person's insanity, we say: 'Send him out.' We also require from superintendents a report on the physical condition, to show whether the patient, on admission, bore marks of having sustained injuries. We have different kinds of reports from different asylums. In case of injury during confinement, the superintendent is required to report the condition of the patient and the date of the injury.

"With regard to transfers from an asylum, practically, we can transfer a patient should we have reason to believe that he would be more benefited by being boarded out or placed in some other institution; but such a thing would only be done after one of our visits of inspection—after we had seen the patient, and it appeared to the inspector that he did not seem to require asylum treatment. The visiting inspector would first discuss the matter with the superintendent. If the discussion was not satisfactory, the man would probably be visited again or examined by some one else."

Family care, or boarding-out, as it is called, is an important feature of the Scotch system, there being, as already indicated, about one fifth of the Scottish insane provided for in this way. On the 1st of January, 1887, those in private dwellings numbered 882 males and 1,388 females, being, as compared with the foregoing year, an increase of 92. Of the total, 130 were private and 2,140 pauper patients. The Deputy Commissioners in visiting these widely scattered patients, many of whom are in remote parts of the country, perform a laborious task unparalleled in the history of lunacy supervision.

It is claimed that the boarding-out of pauper patients in the residences of small farmers affords a convenient outlet for the over-accumulation of chronic cases in asylums, besides conferring upon the insane the great benefits of liberty, air, and exercise. The weekly cost per capita has been estimated at about four shillings below the average asylum cost. A low rate of maintenance in institutions tends, however, to lessen the number of patients boarded out.

The method of placing the insane in families and the supervision exercised by the General Board of Lunacy over those boarded out was thus described by Commissioner Sibbald:

"There are only two or three places in this country where you can see a considerable aggregation of patients. We do not consider grouping in large numbers desirable, but prefer that patients should be scattered. One of these groups is at Kennoway village in Fifeshire; there are others at Balfron and Gartmore in the counties of Dumbarton and Perth. At Kennoway, every thing has been done to prevent aggregation increasing; but it has increased in spite of our efforts. Each family is examined as to its suitability for the reception of cases. For the particular class to which it is adapted in this country, the system works well. The Glasgow parochial authorities are among those who take great interest in the boarding-out of pauper lunatics. They send periodically a committee of their number for independent paid visitation, and take a personal interest in the condition of the lunatics.

"Our general principle is, that, as far as possible, we should have a system which will provide for each case according to its requirements. If a person requires an expensive asylum—one with elaborate structural arrangements, and where a great deal of attention would be bestowed,—he should be sent there. If patients are fallen into a condition that makes them easily managed, and if they do not require very special arrangements, but, at the same time, are not suitable for a private family, we try to place them in an institution where their wants will be supplied. Those who do not come under either category, we endeavor to find suitable families for. As a rule, patients boarded out do not include acute cases. It is an accident if such a case is included, or it is when the family are the natural guardians of the patient. For instance, a man is taken ill, he becomes a lunatic, requires to be aided from public sources, and must be attended to. His family desire our permission to keep him in a private house. If the case promises to be a short one, he is not sent to an asylum, but remains with his friends. Every one who receives public assistance must be reported. If a lunatic is boarded in a family other than his own, he must be reported; but he can remain in his own house for a year, if he does not require special control. If friends require to exercise control to deprive him of ordinary liberty, they can do so, even in his own house, only for twelve months; thereafter he must be reported, and they must submit to whatever we think is desirable. Every person among the insane poor who is boarded out is visited four times a year by the medical officer of the parish, and twice a year by the Inspector of Poor of the parish. These visits are all recorded in a book kept by the person to whom the house belongs, and in the record of the visit there is practically a statement including several details; such as, that the patient is satisfactorily provided for; that he has every thing he requires; and that the conduct of the guardian appears to be suitable. This book contains the six entries every year. It is seen by our Deputy Commissioner when he makes his visit; and if the entries are not made, he takes note of the fact and reports it. We immediately inquire into the reason, and take what steps seem desirable to put the thing right. Not a month passes but some one has to be taken away from a family because of improper treatment. When we hear of such cases we order removal. The District Lunacy Board has nothing to do with this part of the work. We deal directly with the parish that makes application to us through its officers."

The correspondence of the Board in connection with this branch of the work is very voluminous. Each case is examined into with great thoroughness, and the reports and papers relating thereto are separately filed. It appeared from the records that, in some cases, as many as six letters a month had been written. In one instance, the correspondence extended through a period of seven years; in another case, that of a boarded-out female, through fifteen years. Our informant continued:

"These reports the Deputy Commissioner takes with him in going his rounds over the country, and they supply him with a history of each case. Some he does not deem it necessary to visit frequently. They are people who are known to enjoy a considerable amount of liberty and who live in comfortable circumstances—where there is absolute certainty that the patient is comfortable and well treated without being subject to unnecessary control. It may be one living much in the gaze of the public, about whom there is no doubt. In such a case, we do not say, 'You are a lunatic, and I come to inspect you.' It does not rest with the Deputy Commissioner, however, to leave off visiting. The case must be reported here, and we may approve or disapprove of the doctor's proposal to discontinue visits. Frequent visitations are made in unsatisfactory cases; but we desire to avoid 'red-tape' in carrying out our arrangements. We strive to have good reason for whatever we do—to have matters properly considered, and to avoid negligent conclusions."

The great pressure on asylum accommodation and the inducements offered by the lower charge for support in families doubtless have much to do with the extension of the boarding-out system in Scotland. The price paid for the maintenance of paupers boarded out is not uniform, but depends on the locality and its proximity to a market. In justification of the extremely low rates paid in some localities, one of the Deputy Commissioners says:

"The Shetland guardian, who undertakes to maintain an unproductive lunatic at the rate of 2s." (say 50 cents) " per week, may, at first sight, seem to be badly treated, as compared with the Lanark guardian, to whom 8s. per week is paid for the maintenance of a similar patient. When, however, it is considered that in the latter county every article supplied for the use of the patient is a readily marketable article, procurable only by exchange for money or marketable labor, and that in the former the habits of the people and the abundant supply of fish for food, render it almost unnecessary for the careful guardian to expend the allowance in any thing but luxuries, the difference is seen to be only an apparent one. Similarly the standard of comfort varies as much between several of the counties as if they were situated in entirely different countries. In estimating, therefore, whether proper provision has been made for the comfort of an insane patient boarded in a private dwelling, it is necessary to employ not a national, but a local standard, to determine whether a lunatic is treated well as compared with other members of society of the same rank inhabiting the same district."

As much interest attaches to the boarding-out system, a personal inspection was made of a considerable number of the cottages in which the insane are received, and it may not be out of place to describe briefly the homes and surroundings of some of those intrusted to family care. Kennoway, a small village near the Fifeshire coast, was the first place visited. It commands a pleasing view over the Firth of Forth, though distant about three miles from the sea. The narrow and irregular streets and the antiquated look of some of the one and two story houses ranged on either side proclaim the primitive character of the place. Here, as in most of the country villages of Scotland, surface drainage prevails. Some of the dwellings have tile and others thatched roofs; and, although a fair degree of cleanliness is maintained, there are evidences on every hand that the inhabitants are unambitious, there being little scope for enterprise. They have, apparently, succumbed to the inevitable and quietly settled down to their cramped surroundings. Trade, which consists largely of hand-loom weaving, has been on the wane for years,—a circumstance which tends to explain the decrease of population and the moderate rental of the houses. Several rent small patches of ground ; and, besides the few trades-people and small shopkeepers necessary to meet the limited requirements of the place, there are resident in the village, the clergyman of the parish, a medical man, a schoolmaster, and an inspector of poor. Sheltered by heath-clad hills from the cold north wind, and secluded, though not distant from the larger centres of population, Kennoway has apparently been well chosen as a settlement for the insane. And now let us glance at the interior of some of its cottages as described in my notes.

(*a*) Here, in a one-story cottage, plainly built of stone, and having a red tile roof, reside an aged married couple—the husband very deaf. The wife, it must be confessed, is not over cleanly in appearance. The boarders are two female lunatics, one of whom is abed, and the other walking about in her room, which is of goodly size and has an open fire-place. On the papered walls are a few pictures, and the furniture comprises a wooden table and one or two chairs. The other and better furnished apartment occupied by the first-mentioned patient contains a bed in a recess, an inclosed closet, an open grate, a carpet on the floor, and mahogany chairs with hair bottoms. Each boarder is received at six shillings per week. The housewife, referring to the patients, says : " They gie verra little trouble, troke aboot, and are able to dae ony little thing for themsels."

(*b*) This is an antiquated two-story-and-garret stone cottage with red tile roof. It is occupied by a widow who boards four female patients. One of these is subject to epileptic fits, and keeps her bed. The other, a woman nearly fifty-three years of age, has been here about seventeen years. She takes her meals in the kitchen, which is of fair size, and has a red brick floor and an open fire-place. In the four other apartments are open fire-places, carpeted floors, hair-bottomed mahogany chairs, and close-curtained wooden bedsteads. Among other objects seen are flowers and dried grasses, pictures on the walls, Scriptural mottoes, etc. One of the patients says : " The mistress is verra kind to ilka body i' the house, an' we help her to keep it in good order." Each of the patients is paid for by the parochial board at the rate of 6*s*. per week.

The buildings, though in good repair, are many of them old and quaint-looking. The one last mentioned has corbie-stepped gables, and above the doorway is a stone panel with a coat-of-arms in *bas relief* surmounted by a crown and bearing date 1712.

(*c*) In this two-story stone building, which is occupied by a widow, there are boarded four female patients. Two of these occupy separate sleeping-rooms up-stairs, and the other two, who are from the Edinburgh asylum, have a room on the lower floor, in which are two beds. The latter are upwards of fifty years of age, and they occasionally promenade about the large garden in the rear of the house or along the not-much-frequented roadway in front. One of them has a disposition to work. Open fire-places are used, and there is an appearance of cleanliness and a fair share of comfort. The patients living up-stairs have their meals separately conveyed to their rooms. They have oatmeal porridge in the morning, beef broth with potatoes for dinner, and tea at night. The housewife assures us, " They 're a' very easy tae pit up wi'."

(*d*) The tenant of this house is a contractor, who owns a pair of horses and a couple of carts. Two male patients are boarded here, one sixty and the other forty-five years of age. They sleep in separate beds in a room up-stairs with low ceiling. It measures 17 x 18 feet, and has an open fire-place. The furnishing is plain, but seemingly comfortable. The patients dine in the kitchen with the contractor and his wife. Breakfast at 8 A.M. consists of porridge and milk ; dinner between 12 and 1 P.M., of broth with beef or home-cured pork ; at 6 P.M. tea is served with bread and butter; about 9 P.M. bread and milk. On certain days, for dinner, rice and milk with bread and potatoes is substituted for broth. One of the boarders has been five years in this house and is spoken of as " a handy sort o' bodie, who gies verra little trouble and does what he 's bidden." The patients are paid for at the rate of 6*s*. 6*d*. each per week. They find plenty of exercise in the garden, and are said to manifest no disposition "to rin awa,' bit jist paidle aboot the street."

(*e*) This is a one-story cottage with red tile roof. A parapet wall with iron top-railing and two iron gates encloses the front plot of ground, which is made attractive with shrubs, fuchsias, climbing roses, and other flowers. In this clean and tidy dwelling, tenanted by an aged married couple—the mistress being about seventy—are three snug-looking apartments. The two female patients occupy separate beds in a parlor, where an open fire burns brightly. The bedding is comfortable, and the old-fashioned wooden bedsteads are curtained. One of the patients has been here nine and the other seven years. The latter, nearly seventy years old, takes her breakfast in bed. Says the housewife: "The tither ane gets up and scrubs the floor and carries water frae the well." The worker thus referred to is about fifty years of age. Their washing, the aged landlady says, is given out.

(*f*) In a one-story cottage of two apartments, tenanted by a maiden lady, are two patients, one of whom is an old woman who has boarded here for the last eighteen years. The landlady, a weakly person and in humble circumstances, keeps a female servant to do the housework. The interior is not over cleanly, though the patients' room seems rather tidier than the rest of the house. The elderly patient sleeps in a bed occupying a recess in the wall. The furnishing is fair. The patients receive their meals together in the kitchen. One of them informs us that she is "quite comfortable an' has jist to gae oot an' tak' a walk doon the lane when she likit."

The patients living under family care are, as has been stated, distributed over all the counties in Scotland, and efforts are made, as far as possible, to prevent any aggregation in localities. Quiet rural districts, however, afford peculiar facilities for boarding the insane, and where such places are peopled by sober and industrious peasants in humble circumstances, willing to receive, at a low rate, patients of docile character, it is not always easy to prevent undue accumulation. The counties in Scotland containing the largest numbers of patients in private dwellings are Fife, Perth, Edinburgh, and Lanark; where it is to the advantage of the small farmer to board one or two of this class. Many are accommodated and their labor utilized among the crofters of Orkney, Shetland, and other sparsely populated districts, where provisions are cheaply raised, and where even the meagre allowance of the parochial authorities is an inducement. Without wearying the reader by recounting cases in which the details are similar in character, only two more illustrations of family care will be given, as witnessed in one of the more remote of these rural districts. The following are taken at random from the notes made :

(*a*) In a plain, two-story, tile-roofed house, are two male lunatics, one of whom, named Peter, is subject to epileptic fits. He has been here seven years. The guardian informs us that the fits to which he is subject come on at intervals of a fortnight, and that the patient is allowed to go about until the attacks abate. He is described as a hard smoker, and when disposed to consume too much tobacco, a portion is withheld. This invariably enrages him ; but, as the guardian remarks, " He soon calms doon again." Charlie, the other, has been in this house for about twelve years, and was formerly boarded with the father of his present host. Neither of the patients ever tried to run away. Both sleep in a good-sized back bedroom, in which are separate iron bedsteads, with bedding clean and comfortable. One of the sheets has been torn somewhat by Peter, and he is reported as sometimes attempting to destroy his clothing. On the wooden mantel are flower vases with dried grasses and numerous little ornaments ; on the papered walls are pictures, including a large portrait of the great poor-law reformer and eminent divine, Dr. Chalmers. In the plainly furnished kitchen, a grate fire with open range sends forth a ruddy glow. On the upper floor is a front room ready for other patients. It is furnished with Brussels carpet, stuffed-bottom mahogany chairs, a book-case, and a large double, iron bedstead with a brass top-rail. There are two-sash windows with Venetian blinds. The stairs leading to this comfortable apartment are of varnished pitch pine, on which is laid a strip of colored oil-cloth. The patients take their meals with the family in the kitchen.

(*b*) Here, in a low, plain stone cottage with red tile roof, an elderly widow retains as boarders two adult female pauper lunatics, who have been in the house about five years ; also a sane girl about ten years old, but boarded out by the poor-law authorities for the past four years. The little girl is out on a "berrying" excursion with other children of the place. Another of the inmates is a sane blind woman, who has been boarded out by the poor-law authorities for the past six years. The kitchen pavement laid with concrete cement and the parlor floor of wood are clean. The furniture is plain, but tolerably comfortable.

The space within this small one-story cottage seems insufficient for the requirements of the inmates. It is a matter of surprise that so many dependents differing so materially in their mental and physical conditions should receive common care under the same cottage roof.

Dr. Fraser, Deputy Commissioner, whose visitations during 1884 extended to 1,073 patients in private dwellings, says:

"Every year there is in my experience a manifest change for the better in the condition of the insane provided for in private dwellings. Any thing like retrogression in the condition of a patient so provided for stirs up active interference, the end of which, if the Board fail to bring back a satisfactory condition, may be a resort to asylum care and treatment. There is, and I fear there always will be, a small number of the boarded-out insane, chiefly under the care of persons bound to them by ties of kinship, whose condition cannot be regarded as satisfactory. In these cases the condition in which the guardians live is sometimes far from being what is desirable, especially as regards cleanliness and comfort; but when there is a kindly treatment founded on natural affection, it is often thought well not to insist on the separation of parents and children, or brothers and sisters, though there may be a slovenliness and want of cleanliness in the surroundings of the family which is greatly regretted, and which may be found to exist in neighboring houses occupied by poor rate-payers.

"As regards the great majority of the insane in private dwellings in my district, I have not the least hesitation in saying that they are adequately and suitably provided for; that their general condition is a satisfactory one; that they enjoy a more rational, normal, and healthy life than any institution could afford them; that they are satisfied with the arrangements made for their care and comfort; and that they would not exchange their domestic and other privileges for any other treatment of which they may have had experience. The evidence of the correctness of these views lies in the low mortality which exists among the boarded-out insane, their good health and good physical condition, the rarity of accidents among them, and their general contentment."

Respecting the adoption of the system elsewhere, the Deputy Commissioner gives it as his opinion that Scotland has no special advantages for the treatment of the insane in private dwellings, but that "other countries possess retired hamlets and villages, clean and tidy cottages, and thrifty, respectable, and matronly housewives, who would make excellent nurses for the insane. The system has been firmly established in Scotland, and is growing there, mainly, if not entirely, because there is faith in it, and earnest efforts have been made to render its working satisfactory."

Deputy Commissioner Lawson, whose visitations in 1884 extended to 879 boarded-out patients, divides the pauper patients visited into four classes.

"In each house, and with the patient before me, I have recorded in my visitation book my impression of the satisfactoriness or unsatisfactoriness of his care and guardianship. In doing so I divided my patients into four classes, under headings indicating the nature of the provision made for them, namely, Bad, Middling, Good, and Very Good. Of course it is impossible to insist, in localities so diverse in their notions of social comfort as are the counties embraced in my district, on a uniform standard of excellence. We must regard lunatics, whether in Shetland or in Roxburghshire, as being under 'Good' conditions who are as well provided for as sane people in the same rank of life in their neighborhood. Under the heading of 'Very Good' I embrace cases in which the amount of care bestowed on the patients and comfort enjoyed by them are conspicuously superior to what their surroundings would have led one to expect. Under the term 'Middling' are grouped patients who, though on the whole sufficiently well provided for, are neither in themselves nor in their surroundings so comfortable as similar patients in the same district. They make up the class in which advice and expostulation are needed, and are generally effective. Under the heading 'Bad' I include patients so unfavorably situated that an immediate change of guardianship would, apart from other conditions, be considered advisable."

In summing up, Deputy Commissioner Lawson finds the condition of 24.3 per cent very good, 60.6 good, 12.2 per cent middling, 2.9 per cent bad.

Of 67 non-pauper patients in private dwellings visited, the same official reports that he " found them, without exception, suitably provided for, and in a manner in keeping with their resources. Many of them are so completely integrated with the families, whether related or unrelated, in which they live, that visitation is more or less formal. In the houses specially licensed for receiving a limited number of private patients and boarders, I found the inmates and lodgers satisfied and comfortable."

The Scotch Lunacy Board defines its policy in respect to the boarding-out system as follows:

" It is sufficient to say that the policy of the Board, of which the boarding-out has been an outcome, has been to discourage the unnecessary or needlessly prolonged removal of pauper lunatics from the position which they would naturally have occupied if they had been sane; and where such removal is required, either for their own welfare or the public interest, to prevent the restrictions and other circumstances of their treatment from interfering more than is necessary with their natural mode of life. With this view we have striven to prevent the unnecessary or inconsiderate removal of patients from their own homes, to encourage as far as possible the abatement of the prison features of asylums, and to stimulate the relegation of patients from asylums to their homes when asylum treatment ceased to be beneficial. In accordance with this, we have encouraged the transference of patients in asylums for whom asylum treatment had become unnecessary to the houses of strangers in their own position in life, but only when no relatives could be found able and willing to take efficient care of them. In this way the boarding-out has had the effect of diminishing the demand for further asylum accommodation, and has permitted a considerable number of pauper lunatics to live in a way little removed in its character from the mode of life which they would have led had they not suffered from insanity."

A considerable number of institutions for the insane were examined in Scotland, but descriptions of only one each of several different types of asylums are here given.

THE BARONY PAROCHIAL ASYLUM, WOODILEE, LENZIE.

This asylum is managed by an unpaid committee of the local Poor-law Board of the Barony parish of Glasgow. It accommodates upwards of five hundred patients, mostly from the city, a large proportion of whom are the acute, suicidal, and dangerous. The Barony parish was the first in Scotland to avail itself, in 1875, of the powers of the Act already referred to, enabling poor-law authorities who had already provided accommodation for their insane, to continue to do so under license of the Commissioners in Lunacy without interference from the district boards. That enactment is now popularly known in Scotland as the Lenzie Act. Under the powers thus conferred upon them, the Barony parochial authorities erected, at a cost of about £300 per bed, one of the best pauper asylums in Great Britain. It occupies a healthful and pleasant country site, ten miles distant from Glasgow and three fourths of a mile from Lenzie. It has attracted much attention on account of the amount of liberty extended to the patients, the absence of locked doors and all kinds of walled enclosures, and the substitution of well-directed employments for mechanical or chemical restraint.

There are three hundred and eighty-two acres of land belonging to the asylum. Twenty acres are reserved for exercise and recreation; the remainder, as also some leased land, making in all upwards of four hundred and fifty acres, is worked under a thorough agricultural system, and affords outdoor employment to a large number of the insane. A considerable number of working patients occupy detached dwellings on the farm lands.

At the entrance to the asylum is a neat porter's lodge occupied by a married attendant and several patients. The asylum buildings, which are of two stories, are artistically designed and advantageously disposed. The grounds are tastefully ornamented with shrubbery and flowers. Nothing in the semblance of wall or enclosure meets the eye, if we except an iron fence at a distance of ninety yards from the main building, and bordering the Edinburgh and Glasgow Railway. This road extends three quarters of a mile through the asylum grounds, and nearly two hundred trains pass over it daily.

The asylum officers consist of a medical superintendent and assistant, a chaplain, clerk to committee, house clerk and steward, assistant steward, matron and assistant matron. There is one attendant to ten patients. The male and female sides of the building are each divided into four distinct sections. Each section contains about sixty beds, and is in charge of a chief attendant, under whom are three subordinates. At the time of my visit, there were in all 476 inmates—241 men and 235 women.

The lower story of the asylum building is set apart for day use; the upper, for night occupation. In the dormitories, the uncarpeted polished pine floors have an appearance approaching elegance. The polish is imparted by using a compound of beeswax and turpentine, which is applied with a soft cloth fastened on a weighted block. The work of polishing is done by patients who cannot otherwise be well employed. The walls are tinted and adorned with pictures. The air is made fresh and pure by opening the windows as soon as the rooms are vacated by the patients. By means of two large open coal fires the dormitories are heated and ventilation is promoted. The windows of the upper apartments have neither iron sashes nor outer guards. This arrangement, approved by Dr. Rutherford, the superintendent, forms part of the free system obtaining throughout the institution. For some of the single rooms, sliding shutters are used. As was found elsewhere in Scotland, the beds were comfortable and bed-clothing abundant. Each bed was furnished with a palliasse, hair mattress, two pairs of heavy blankets, sheets and coverlet, hair bolster, and feather pillows. The bedding is carefully attended to, and the sheets and blankets for each division have a distinguishing stripe. The strong wooden bedsteads have elastic slats and substantial head and foot boards. They are taken to pieces for renovation once a fortnight. The associated dormitories have each two small bureaus and chests of drawers, a double breadth of three-ply carpet in the centre of the room, a rug by the side of each bed, besides minor articles which add to the comfort and cheerful appearance of the room. Each associated dormitory has eleven beds.

In the fourth division there are eight single rooms. Altogether there are in the institution one hundred and fourteen single apartments. Two strong rooms on either side have double doors, and windows effectually protected by sliding wooden blinds.

The infirmary wards are even more liberally furnished than the other apartments. The furniture includes Brussels carpets, easy and invalid mahogany chairs, centre tables, bookcases, etc. In the rooms for convalescent patients, the windows have ordinary wooden sash, the panes measuring 10 x 16 inches. Epileptics are separately cared for and closely supervised in the infirmary department. Here are large unscreened open fire-places. This department has an attractive aviary, in which the patients seemed much interested. Suicidal patients are assigned to the various wards, and over them a special watch is maintained.

The bath-rooms are conveniently arranged, and furnished with mirrors, combs, brushes, etc. Water is pumped to reservoirs in the towers of the building, whence it flows to every part of the establishment. A dial on the lower floor indicates the quantity of water stored.

Two sewing or day rooms that I entered on the ground-floor are likewise cheerful and well-furnished apartments, each having a work-table, piano, comfortable lounges, flower vases, and other articles of use and decoration. The open fire-places are made more attractive by bright-colored tiles. Pleasing outlooks are obtained from the wide bay-windows. A considerable number of women patients were here at work, apparently contented and happy.

There is a large, tastefully decorated hall with raised platform on which was a piano. The hall is appropriately furnished and is used for entertainments, including weekly dances, which are exceedingly popular with the inmates. A band composed of attendants and patients plays every Saturday afternoon, either in the hall or out-of-doors, according to the weather. The asylum is well supplied with reading matter, consisting of books and cheap publications.

In the dining-hall I saw about 225 men and nearly as many women at dinner. The tables, covered with snow-white linen, were ranged in three rows or divisions. There were chairs for eight persons at each table. On one side of the hall were the men, and on the opposite side, the women; while the intermediate centre space was occupied by the attendants. The table furnishing, except for the suicidal, embraced a variety of articles, including casters, water-bottles, ordinary spoons, knives and forks, soup and dinner plates, bright metallic soup ladles, tureens, and salt-cellars. Each patient was provided with a table napkin. The meal comprised Scotch broth and dumplings, with meat and a plentiful supply of vegetables and other nourishing food. Half an hour is allowed for the patients' dinner; then follows that of the attendants, which is partaken of in the presence of the patients and while they remain sitting at table. Both have the same kind of food, except on two days of the week, when the attendants have soup and fish for dinner. The insane are not permitted to carry food from the table, and care is taken to remove all the dishes and knives and forks before they leave. The windows of the capacious dining-room are large, and have single panes to each sash. It was stated that but one of these had ever been broken by the inmates. The superintendent said he preferred bringing as many of the insane as possible into the dining-hall; for, he added: "The most difficult as well as the most important part of asylum work is for the attendant to learn the peculiarities of his patients." The sick, however, as well as those who cannot be trusted in the dining-hall, have their meals carried to them. Epileptics dine with the other inmates. Two had attacks of epilepsy during the meal.

The kitchen is of one story, and has a large central hood or funnel for carrying off the steam from the food while cooking. Thirty brightly polished tea and coffee urns were observed in their appropriate places. The floor was of tile, enhancing the neat appearance of the apartment, which, with the store-room adjoining, was orderly arranged and carefully kept. Throughout the asylum, every article of brass or steel, including knobs, faucets, and spittoons, was burnished. In the halls and corridors were laid breadths of linoleum with bright colored borders.

Dr. Rutherford, as already intimated, is one of the foremost advocates of non-restraint, and also of healthful exercise and employment. It was stated that there were no strait-jackets nor restraining dresses — nothing involving mechanical restraint within the institution. If violent, a patient is walked about until he calms down ; and if very violent, he is placed in charge of two or more special attendants. Notwithstanding the freedom allowed, few escape who are not brought back the same day.

No fewer than one hundred and forty-eight men were employed on the farm and grounds. In very wet weather, these men are placed at work under shelter, but a slight shower of rain is not deemed hurtful. Patches of ground alike in size, and an equal number of patients are assigned to each of the different attendants working in the extensive gardens, who are thus stimulated to compete for the largest product. The ground for vegetables is cultivated chiefly with the spade. In managing the laboring force, some tact appeared to be exercised. As we approached the stone-quarry where a small group were employed, an Irishman sprang upon a mound of earth and broke into loud speech, accompanied by violent gesticulation. No one paying the slightest attention to him, he soon took up his tools and went to work. Seventeen men were occupied as mechanics in the different workshops for various crafts, including shoemaking, carpentering, tailoring, tinsmith work, gas-fitting, and blacksmith work. About forty weakly patients and epileptics were excluded from the working groups. All, however, not unfit to labor, are obliged to go with the working parties, and it is said that example and precept seldom fail to induce even the reluctant to work, although, nothing in the character of punishment for idleness is resorted to. The farm stock at Woodilee included goodly sized herds of cattle and flocks of sheep.

The sewage is utilized on the land. The original plan was to have only earth-closets; but these were discarded, owing to difficulty in keeping them in order. The closets are in projections from the main divisions. No sewers enter the asylum buildings, but terminate just outside the walls, where they are trapped near the junction of a ventilating flue. All interior pipes are trapped within the building, and are carried through the wall into the sewer. The sewer-pipes, laid below frost depth, extend to a large covered vat some distance from the asylum. The sewage is there diluted and forced by steam pumps through iron pipes to hydrants at elevated points on the farm, from which it is distributed over the land by means of movable troughs. The fields thus treated were much improved. The main iron sewer-pipe measures five inches, and the others three inches in diameter. Intermediate flues act as a protection against sewer-gas. The system is highly approved by the Superintendent, who thinks that the immediate distribution of the sewage obviates any danger to health.

In affording employment, the laundry is to the women what the farm is to the men. Machinery has been discarded for washing by hand. The clothes are dried in the open air when the weather permits. Out of a total of one hundred and forty-five women workers, about thirty were employed at laundry work; sixteen were assisting in wards; thirteen in the kitchen; two were serving as housemaids; fifty-nine were doing needlework; and twenty-six were knitting.

The inmates are variously and comfortably attired. The night-dresses in use in the several divisions are, like the bed linen, distinguished by different colored stripes. The male patients generally wear stout corduroy or moleskin trousers. Male attendants wear uniform caps displaying the asylum badge. Female attendants are dressed in black, which contrasts well with their clean white ruffed caps, white linen cuffs, and white aprons.

Night attendants are governed by rules similar to those in many other British asylums. It is the first duty of the night attendant to ascertain from the head attendant all names of patients who require particular attention in the administering of medicine or special comforts, and of those disposed to suicide. Epileptic and suicidal patients must be visited every hour during the night, and every patient must be looked at when the night attendant enters upon his watch at eight o'clock and again at ten o'clock. Each dormitory must be visited at regular intervals four times during the night. All epileptic and paralytic patients must be examined at each of these visits, and if the linen is wet or soiled, it must be changed at once. The night attendants are further required to know all patients of irregular habits. These must be periodically awakened to attend to the calls of nature. Some of them must be called every two hours; others only at midnight or possibly again at two A.M. It is explained to the attendants, that, with the exception of epileptic patients in a fit, these precautions, if strictly observed, should entirely obviate all wet beds in ordinary cases of chronic mental disease, and conduce to cleanly habits by day. Night attendants are required to make a particular record of all the incidents that occur, including the manner in which newly admitted patients and the sick have passed the night. To restless patients, the night attendants are urged to give particular attention. They are told that a kind word will often bring the patient quiet sleep, while its omission may cause him to pass a noisy, restless night with consequent excitement next day. The attendant is also told that the timid should have their fears allayed by frequent friendly visits, in the expectation that sound sleep may be secured to them through the consciousness of the attendant's presence. Night attendants are charged with the duty of seeing that there is nothing done after the inmates have retired to disturb their rest. At 5.30 A.M. the night attendants call the day attendants. Until breakfast-time the former aid in dressing and washing the patients.

At the time of my visit, the daily routine observed here was as follows: The morning call is made at 5.30. Day attendants superintend the washing and dressing of patients, and see that their beds are turned down and exposed to the air. At 6.30 artisans and selected farm and garden patients begin work, also selected kitchen and laundry patients. At 7.30 the attendants take breakfast. At 8 o'clock the patients have their breakfast. At 8.30 there is morning prayer in the chapel. At 9 o'clock the working male patients are inspected in the court, and they afterwards go out to labor. At the same hour the kitchen, laundry, and sewing women are set at work. At 10.30 o'clock a medical visitation to the wards is made. From 10 to 11 extras for the sick are issued from the kitchen or stores. At 12 M. medicines are dispensed and given to the head attendant. At 1 P.M. working patients return to the wards and make themselves tidy for dinner. At 1.20 o'clock is the patients' dinner, and the attendants dine twenty minutes later. At 2 the working patients return to work. At 4 o'clock extras for the sick are issued and tea is given to laundry patients. At 6 o'clock in summer working patients return to their wards. At 6.20 patients have their tea or supper; at 6.40 attendants take their supper. From 7.30 to 8, patients go to bed; all shutters and doors are locked, clothing removed from rooms, and lights extinguished. A second medical visit is made about this hour. At 8, night attendants enter upon their duties. At 10 o'clock all attendants except those on duty go to bed and the lights are extinguished. The rules set forth that the hours of outdoor exercise shall depend upon the weather and season of the year, and that two hours daily should be spent in the open air by the non-working male and all the female patients.

Two corridors inclosed with glazed sash extend from each wing of the asylum rearward to the chapel on the lawn. These cheerful promenades, lined with fragrant flowers, have the appearance of beautiful conservatories, and afford pleasant walks in the most disagreeable weather. In the chapel—a tastefully built stone edifice with attractive stained-glass windows—services are conducted every Sunday by a local pastor. The men and women occupy different sides of the building. There are religious exercises here every morning, consisting of singing, reading, and prayer.

An air of industry pervaded the institution, and thorough administration was everywhere apparent. In a walk about the grounds with the Superintendent, a considerable force of men were seen grading, planting, and making other improvements. Others were engaged on the lawns and grounds, working with little or no manifestation of excitement. Successful efforts were made to get the patients interested in their work. Both patients and attendants labored together, the latter by example stimulating and encouraging the former, while taking care to prevent undue exertion on the part of the patients.

This asylum was examined with peculiar interest, for the reason that it has the reputation, beyond any other of its kind in Great Britain, of making practical application of the most advanced theories respecting non-restraint and personal freedom. Of this feature of the asylum and its results, Dr. Rutherford says: " The diminution of restraint is beneficial, inasmuch as it renders the patients more contented, and makes them practise self-restraint more than they feel disposed to do when others seem to be doing it for them. Besides, there is nothing more irritating than forced restraint. Many, inside and outside the asylum, go about and work unattended. Freedom does not increase the number of suicides, which have been of very rare occurrence here. Conflicts with attendants are also rare ; and when they do happen, it is generally indoors. It is the rule that attendants, under no pretext whatever, shall leave or lose sight of a patient without placing him under the care of another attendant, and the rules are very strict respecting rude treatment. If it becomes necessary to seclude a refractory patient, the fact must be immediately reported through the head attendant to the Medical Superintendent. Under this system, it cannot be said that the number of attendants is actually or necessarily increased. The excess, if any, is of that class of attendants who, by their work, earn their wages, and are, as it were, extra ; such, for example, as artisans and gardeners. These attendants work with the patients, and are generally the best workers in the party. The patients imitate and follow the attendant, only he must be able to handle a pick and shovel better than they, else they will work very little—they will simply play at work. Here, it is easy to get such attendants. The responsibility of the superintendent is not increased. Greater vigilance may possibly be needed on the part of attendants to prevent escapes ; though under the system of enlarged liberty more escapes do not occur than frequently happened in the old lock-up institutions."

In respect to stimulating beverages and medical treatment, Dr. Rutherford says : " There is a diminishing use of wine and spirits. Beer is not used by the patients as an article of diet. Forty gallons of new milk are consumed daily. Sedatives and narcotics are not used except where sleep cannot be otherwise obtained. Only twelve out of the total received draughts for this purpose last night. Insanity is essentially a disease of diminished vitality, and when present, the system demands invigorating treatment. Experience proves that there is nothing so invigorating as active outdoor employment and abundance of fresh air."

Without attempting to decide to what extent the radical principles put in practice here are worthy of general adoption, it was evident that there was a remarkable degree of contentment and cheerful activity. The visit, on the whole, left the impression that the Barony Parochial Asylum was conducted with great energy and ability, and that it is entitled to the high position it holds among British institutions.

The number of patients registered on the books of the asylum September 3, 1887, was 531. They were distributed as follows: Within the main building, 473; residing at the Muckroft farm, 12 men and 2 women; at the new Farm Steading, 26 men and 2 women; at Fauldhead, 8 women; and at the Gate Lodge, 6 men. Two women were absent on pass. The net per-capita cost of maintenance, as officially given, was 8s. 9d. per week, for the financial year ending May 14, 1887. The force of attendants has been proportionately enlarged with the increase of numbers.

Since my visit, Dr. Rutherford has been called to assume charge of the Crichton Royal Institution at Dumfries, and has been succeeded by Dr. Robert Blair. The Lunacy Board in reporting upon the institution in 1887, said: "The ample means of industrial occupation afforded by the arrangements of this asylum continue to be judiciously taken advantage of. The patients were free from complaint, noise, or excitement, and the establishment was found in excellent order throughout."

MID-LOTHIAN AND PEEBLES DISTRICT ASYLUM.

This asylum is located at Rosewell, a few miles from Edinburgh, and is easily accessible by rail. Its site is seven hundred feet above the level of the sea, and commands extensive and pleasing views of the adjacent country. The asylum was opened in 1874. It is managed by an unpaid district board. The Medical Superintendent controls the selection of the subordinate staff.

In the original estate there were only about forty acres. The disadvantage of having so limited an area led to the additional purchase of a small farm ; but even with this the amount of land is found to be insufficient. The plain and new-looking two-story buildings of cut stone, having accommodation for about two hundred patients, were erected at a cost of £50,000. The surrounding grounds are well kept, and the terraced lawns are adorned with flowers and shrubbery. The institution has, in addition to the main building, two convenient and comfortable cottages for attendants and patients.

This asylum provides for the requirements of the landward district of Mid-Lothian and Peebles. So far, it has had accommodation to spare after meeting the wants of the district rated for its support, and accordingly a number of what are termed "out-county" patients are received. About one fifth of the inmates are private patients, who are charged, if residing within the district, 11$s.$ 6½$d.$ per week, and if without, 14$s.$ "Out-county" pauper patients are received at the rate of 12$s.$ 3⅜$d.$ per week. The rate for pauper patients within the district is officially given for the year ending May 14, 1887, as 10$s.$ 4⅜$d.$ This includes provisions, household requisites, medicines, clothing, salaries and wages, furniture and furnishings, and incidental expenses.

The upper floor is set apart chiefly for dormitory purposes. On the lower floor are the sitting or day rooms, dining-hall, principal lavatory, a few single rooms, and other apartments for day use. The dormitories vary in size, containing from fifteen to thirty beds each, and they, as well as some of the single rooms, appeared to have insufficient air space. Flowering plants and ferns were about the windows, which have plain wooden sashes with panes 8 x 10 inches. The windows are ungrated, and are generally without either shutters or blinds. The sash can be lowered at the top and raised at the bottom about six inches. Rugs and breadths of Brussels carpet were laid on the waxed floors. The tinted walls had bright paper borders. The bedsteads were mostly of wood, with head and foot boards. The mattresses were in three sections. The bedding was abundant and comfortable. Some woven-wire mattresses that had been received on trial were approved. A lavatory examined on the upper floor, was well supplied with towels, brushes, combs, and mirrors.

In the dormitories, even in the single rooms, and in the apartments generally, are fire-places. These are supplemented by a complete system of steam-heating with direct radiation. Ventilating flues are in the walls. The lighting is by gas. In some of the apartments the gas-jets have frosted globes. Altogether, there are from thirty to forty single rooms. In all of those on the upper floor there are concealed adjustable shutters, by means of which the windows may be effectually protected. It was said that these were rarely used.

One of the day or sitting rooms that I inspected was a pleasant apartment with a broad and deep bay-window ornamented with lambrequins. The walls were tinted and hung with pictures; the wood casings polished and varnished, and on the floor were cocoa-fibre rugs with colored borders, and a broad piece of linoleum extending through the centre of the room. The chairs and lounges were upholstered with enamelled canvas. A billiard table afforded a kind of recreation that was much appreciated by the patients. Geraniums and other flowering plants were here displayed to advantage.

Adjoining this was a goodly sized lavatory with floor of colored tile, highly polished brass-work, abundance of clean towels, and mirrors on the walls of three sides of the apartment. The infirmary wards were very comfortably furnished. In the halls and corridors the walls were suitably colored, the ceilings plainly decorated, the windows curtained, and along the waxed floor was a breadth of bright linoleum with a brass protecting border screwed firmly to the floor.

The men and women have their meals in a hall which had tinted and wainscotted walls, and was furnished with ordinary chairs, tables with linen covers, crockery plates, knives, forks, and spoons. One end of the room was apportioned to male, the other to female patients, who enter by different doors. They were served by attendants, one of whom presided at the head of each table. The dinner consisted of broth, beef, potatoes, and bread.

The chapel, which is also used for entertainments, has ornamental interior roof-work, colored glass windows, and, considering its purposes, is elegantly fitted up. A small organ and a piano were on the platform. This was covered with Brussels carpet, and a breadth of the same was laid along the passages, the clean waxed floor showing at the sides. The seats or settees were comfortable and were conveniently arranged.

At one time much complaint was made about dampness, which had become apparent on the walls of the day-rooms, dormitories, and corridors of both floors of the asylum building. To obviate this serious difficulty, an experienced builder recommended, among other things, that the exposed portions of the stone walls be painted with sugar-of-lead and oil, and that all imperfect stones be taken out and perfect ones substituted. His suggestions were acted upon, and the result was satisfactory.

There were neither close nor padded rooms, and it was said that the principle of non-restraint was observed as far as possible. In certain cases, as, for example, soothing a patient when in a state of great excitement, some such preparation as hyoscyamine, or perhaps a subcutaneous injection of some kind, was resorted to. Muffs were hardly ever, and locked dresses never, used. The attendants at this asylum were in the proportion of one to twelve patients, and their pay was periodically increased according to merit and length of service.

Fifty per cent of the men in suitable weather are generally occupied on the farm or in improving the grounds. These appear more spacious from the absence of walled airing-courts. The women are extensively employed at laundry work, sewing, knitting, and house-cleaning. The patients were not uniformly dressed. Their clothing was substantial and comfortable. In winter, the men who go out to work wear heavy plaids and thick woollen gloves. Care is taken to see that they have water-tight shoes in good order, and also that their strong fustian clothes are in satisfactory repair. In stormy weather, labor is mainly confined to the tailor's, shoemaker's, joiner's, and engineer's workshops. Here, in the country, in comparative seclusion and under proper supervision, a commendable measure of freedom is allowed.

THE ROYAL EDINBURGH ASYLUM—MORNINGSIDE.

Two miles southwesterly from Edinburgh, and just a little removed from the activities of the picturesque metropolis of Scotland, is situated the Royal Edinburgh Asylum. A high place is deservedly assigned to this institution, which, like every other asylum for the insane in Scotland, is managed by an unpaid board. It was organized as a private or chartered corporation, but it has come to serve also the purpose of a district asylum. It is under the efficient direction of Dr. T. S. Clouston, one of the foremost of British alienists. The asylum comprises three separate branches, called East House, West House, and Craig House. A little apart from these, and separated from one another, are several small asylum cottages partly hidden by foliage and surrounded by green lawns and blooming parterres. The country prospect, with a background of heath-clad hills, is pleasing and extensive.

The number of inmates in this, the largest asylum in Scotland, averages about eight hundred. The West House is occupied by about five hundred pauper patients, and private patients who pay low rates for board. East House, Craig House, and Myreside Cottage are occupied by patients paying higher rates. In addition to the buildings referred to, the directors provide, during the summer months, a seaside villa at Cockenzie, a fishing village ten or twelve miles east of Morningside. The rate charged for the support of pauper patients from the district is £33 10s. per annum. The minimum rate for private indigent patients is the same. Other private patients are charged £84 and upwards per annum.

This asylum has a charity fund of about £45,000, created by voluntary contributions, the income of which is expended for the relief of deserving patients in reduced circumstances.

About £14,000 have recently been added to this fund by the generous bequest of Mrs. Elizabeth Bevan, daughter of Dr. Andrew Duncan, who was one of the first Physicians in Ordinary to the asylum.

The West House is a plain-looking but substantially built brick structure standing a little apart from East House. At the entrance gate we met, in charge of attendants, about a score of women comfortably and variously attired, who were quietly taking their daily walk. Near the shaded avenue, a considerable number of patients were seen at work with attendants. The well-kept garden, neatly trimmed lawn, carefully raked gravelled walks, and a profusion of flowers and shrubbery in their summer splendor, denoted well-directed labor. Looking over the spacious and highly improved grounds, the eye failed to detect any sign of irksome restriction in the form of interior walls or other barriers. Connected with the asylum proper were about one hundred and forty acres of land, portions of which were set apart for bowling, lawn tennis, cricket, and other outdoor games. The cricketers, organized as a club, were playing a friendly match with a team from the city.

On the day of my visit, the dinner for the pauper class consisted of hotch-potch (a wholesome vegetable soup), a liberal allowance of corned beef, with abundance of potatoes and bread. The poorer class of paying patients received, instead of corned meat, roast beef with cabbage and other vegetables. All were allowed ale, but some, by special order, received milk. Separate dining-halls are provided for the pauper and paying patients. At meal time an attendant is stationed at either end of each table to apportion and serve the food. The hours for regular meals, are—breakfast at eight A.M., dinner at two P.M., and tea at six P.M.

Pauper inmates have three meals; non-pauper inmates

are allowed a supper in addition. Attendants take their meals in their own rooms half an hour before the patients have theirs. Most of the food is prepared in a general kitchen. Small diet kitchens adjoin the apartments for the sick. In cold weather the plates are warmed before serving the meals. In the culinary department the head female cook is assisted by three attendants and about eighteen patients. The large dining-hall has frescoed ceiling and varnished wooden rafters, and is lighted from the roof. Here are several rows of tables with white linen covers. The men and women occupy opposite sides of the hall. Plates of earthen-ware, horn spoons, knives and forks, pepper-boxes, salt-cellars, mustard pots, and water-bottles were on each table. The knives have only about one and a half inch of cutting edge. The floor is of plain wood, and it was cleanly scrubbed. At one end of this dining-hall is a raised platform, the apartment being occasionally used for concerts and other entertainments. There is, in addition, a large recreation hall set apart solely for entertainments. This hall is handsomely decorated, and the side lights have bright-colored shades. The system of heating is by hot water.

The dining-hall for poorer paying patients accommodates about one hundred at five tables, each of which, with its white linen cloth, was arranged to seat twenty persons. The chairs were of cherry wood, which was highly polished. Knives and forks were used. Earthen plates, casters, and water-bottles were on the different tables. A mahogany sideboard and other homelike articles improved the appearance of the hall, which was lighted by gaseliers and side lights. Throughout the house one is pleased with the general distribution of comfortable furniture, the arrangements for diffusing cheerful light, and the effect of brilliant coloring.

Patients, on leaving for work, pass through what is called

the "shoe-house," against the walls of which are ribbed shoe-presses with separate divisions for keeping boots and shoes, one division being assigned to each person. The workers are divided into parties of ten or twelve, each group being in charge of a working attendant. On their return from labor, the patients assemble in a court with side shelters. The tools are restored to their places, and something like military order is observed. Labor is found to be an effective aid in treatment, and the weather and circumstances permitting, every male patient is, shortly after his reception, set at work in the garden. During the first three days he is under close observation. If a patient proves troublesome and is disinclined to work, or disposed to hinder others, he is placed under the special charge of an attendant. The whole industrial department is intelligently supervised. The carpenter's shop is a well-lighted room, in which were employed two paid joiners and four working patients in making and repairing tables, chairs, wardrobes, wash-stands, mantel-pieces, boxes, etc. Paid workers were also observed in other places as follows: In the tailor's shop, employed in making clothing and bedding, and in the shoemaker's shop, one attendant to every four patients; in the upholsterer's shop, engaged in making straw and hair mattresses, feather pillows, covering chairs, and upholstering old furniture, one attendant to three patients; in the tin-smith and plumber's shop, two to the same number of patients; and in the bakehouse, two to one patient. In the smithy, three patients worked as blacksmiths. In another division about twenty patients in charge of two working attendants were picking hair for mattresses and pillows.

In all of the mechanical departments ordinary tools are used. Implements are carefully kept and must be accounted for by those using them. In the printing-office were three working patients who not only print the reports and blank forms used in the institution, but also a monthly periodical called the "Morningside Mirror," to which the inmates contribute. The windows of the workshops have ordinary wooden sash, the panes measuring 6 x 8 inches. Notwithstanding the freedom here permitted and the use of numerous tools, it was asserted that personal injuries are of rare occurrence. The laundry gives employment to a considerable number of women, and is a well-organized department. For sanitary reasons, the clothes are dried in the open air when the weather permits.

Stores are orderly arranged, and are given out only on approved requisitions from the several departments.

In the older portions of the asylum, there are, in some cases, from forty to fifty patients in an associated dormitory; in the newer parts, but from six to twenty. There is a considerable number of single rooms. In some of those occupied by violent patients, inside sliding shutters are provided for use at night. Some of the bed-frames have adjustable bottoms of strong canvas, and each is furnished with a hair mattress. In addition to other means of heating, open fires are general. Dr. Clouston prefers pitch pine for flooring, and many of the floors are of this material. They are carefully waxed, and present a glossy and even elegant appearance. In the lavatories there are highly polished heavy copper basins. The baths are furnished with every convenience. The closets are in structures separate from the wards, but connected by passage-ways. Except at a portion of West House, the open-door system generally prevails, the doors remaining unfastened in summer until ten P.M.

Walled enclosures or airing-courts have been abolished at this institution. One of the first things Dr. Clouston did after his appointment in 1873 was to remove the iron gratings from the windows of the asylum and take down from about the airing-courts the then existing high walls. The marks where these unsightly barriers terminated against the East House may still be seen. In their place has been substituted, in some instances, a low stone wall surmounted by an ornamental balustrade, which does not obstruct the range of vision. One such limits the recreation grounds for women at the East House.

In the pauper department the means of restraint consists of gloves and the old strait-jacket or "polka"; but these are so rarely used, it was asserted, that, practically speaking, there was no mechanical restraint. Sometimes seclusion is resorted to in a light room; at other times in a dark apartment. In each division of the West House there is a canvas-padded room,—one for men and the other for women. These are warmed as occasion requires, and may be lighted from the ceiling. At the East House were seen two male patients who had been in close charge of attendants for the previous four days. One was comparatively quiet; the other exceedingly boisterous and wild, running and shouting around the green, the watchful guardian all the while keeping close by and permitting him to spend his superfluous energy without injuring himself or others.

The training of the attendants is manifestly of a high order, developing patience and even tenderness. Before entering active service, they are instructed in the duties of the various departments. For cases requiring temporary attention at their own homes, a trained attendant is sometimes sent outside the institution, but only for short periods, and at fixed rates. Patients are brought to the asylum from the Edinburgh district free of charge; from a greater distance the rate is ten shillings per day for an attendant, with travelling expenses. At the time of my visit, a party was seen starting from the door of the East House for the seaside resort at Cockenzie. It consisted of two male and ten female patients, who were comfortably seated in a large open carriage, and accompanied by one male and one female attendant, besides the driver. From their dress and general appearance, they might have been taken for a party of ordinary pleasure-seekers.

The East House, as already stated, is occupied entirely by paying patients. In passing through its various apartments, one is pleased to note the comfortable and homelike furnishings, of which there is a great variety. There are sofas, lounges, rocking and easy chairs, mirrors, draped curtains, chair-beds for attendants, Brussels carpets, carpet rugs, solid mahogany dressing-cases and chests of drawers, towel rails, basins, stands, and cabinets. There are also open fireplaces. On the men's side, are rooms for billiards and other games, and pleasant reading-rooms with well-stocked book-cases. The dining-rooms at East House are small. In one fitted up in oak and appropriately decorated, sixteen patients were assigned to a table.

In the women's department, the furnishings and fittings are also tasteful and comfortable, with possibly a greater profusion of homelike articles. Green Venetian blinds, white lace curtains, mantel-pieces, mirrors in gilded frames, gilt time-pieces under glass shades, carved chimney pieces, paintings, flowers, birds, and many ornamental articles were here seen. The windows have pleasant outlooks. In chilly weather bright, copper gas-stoves are used to temper the atmosphere. The whole department was in charge of a matron, and bespoke careful management as well as good taste. In few other asylums visited did the inmates approach so near the appearance of sane people in home-life.

In regard to permitting the insane to act out their peculiar fancies, Dr. Clouston says: "We now discourage and keep down those outward expressions of insane delusions that used to give a lunatic asylum its most striking character. The monarchs crowned with straw, the duchesses in gaudy spangles, the field-marshals with grotesque imitations of military uniform, that could be seen in any asylum of old, have disappeared in outward semblance, just as in the world at large you may meet an empress and not know her. The public opinion against individuality is as strong now in asylums as in society. If the man with ten millions of money, who is the rightful heir to the throne, affixes the top of a soda-water bottle to the front of his cap, as a faint symbol of his position, it is at once unfastened. If the princess, who is the greatest beauty in Europe, bedecks herself with brilliant bits of incongruous ribbon, it is quietly removed at night. The insane man, like his sane brother, soon adapts himself to his circumstances, and submits to rule and public opinion."

Since my visit, some very desirable changes have been wrought in Morningside, enabling the Superintendent to put in practice some of his excellent theories respecting asylum management. A description of the important improvements effected in the West House, may be best set forth in his own words.

"After the female wards Nos. 1, 2, and 3 in the West House were finished and painted and refurnished to some extent, the two West House dining-halls were repainted and decorated in an extremely tasteful and cheerful style. The comfort of management, and the advantage to the patients, of all that has been done to improve the house, is incalculable. We are trying to individualize and classify the patients in the ward for the admission of the rate-paid women, who are not sick or very weak in body, on the principles adopted for the care and cure of the sick and weak in the hospital wards. That implies, first of all, a large staff of attendants there, and good attendants. This ward consists of three sections, with two large corridors. There is much elbow-room everywhere, the average floor-space to each patient being sixty superficial feet, so that there is no huddling together of patients. The first section is a workroom under the charge of a special attendant, where every suitable patient is at once tried, and, if possible, got to do something, thereby the mind being diverted from morbid thoughts, while self-control is practised and tidiness of dress cultivated. The work-hours are short, with walking in the grounds between. The next division of the ward is a large cheerful saloon with a corridor ten feet wide, where the patients who are admitted in a more excited state, or the very suicidal, are placed. Such patients are put under the individual charge of one or two attendants, whose duties are confined to them alone. Individualization is the key-note in this section. Lastly, there is a smaller sitting-room, with a corridor attached, for a few who are safe to themselves and others, and do not need so much supervision, and where, in fact, the sort of patients are apt to be allowed to sit who from delusions are specially intolerant of too minute supervision (for there is an immense variety in the symptoms of mental disease that has to be provided for), and are best left alone, for a time at least. The sleeping accommodation is also varied, consisting of two large dormitories, in one of which every new patient sleeps for the first few nights at least, and where a night-nurse sits up and makes a written report to me every morning for the first fortnight, as to how every new patient has slept. In the next dormitory four attendants sleep near the new patients. Then there are two dormitories for four patients each, and two with three beds, and twelve single bedrooms. The whole ward accommodates forty-three patients and their seven attendants. I look on it as being of the utmost importance how the new patients are treated and cared for, what first impressions are made on their minds, and how their cases are gone into."

In a detached two-story cottage, formerly the residence of the superintendent, are accommodated four ladies who provide their own attendants. The immediate surroundings of the cottage are beautified by rockeries, flowers, shrubbery, and well-kept walks. The windows have each two large panes of plate glass, and they are provided with inside shutters, green Venetian blinds, and white lace curtains. The sashes may be raised or lowered a few inches for ventilation.

Myreside Cottage, a cozy-looking, old-fashioned farmhouse, with vine-covered porch and thatched roof, is reached after a short walk from the East and West houses. It was successfully converted from a private dwelling into a home for male patients. At the time of my visit there resided here nine persons in charge of two male attendants, who said their duties were not difficult. The doors were unlocked. A woman cook and a scullery maid are here constantly employed. A room up-stairs is used as a dining and sitting room. Open fire-places assist ventilation. No gratings appeared on the windows, nor were there any visible barriers; indeed, the house did not differ from an ordinary rustic dwelling. At one end of the cottage, in a small green-house, were pots of bright geraniums, carnations, calceolaria, ferns, etc. Fruit-trees were trained closely to the garden walls. In front there is a small lawn planted with flowers and ornamental shrubbery.

As one proceeds along a wooded slope, about a mile from the main buildings an ancient and secluded mansion, surrounded by tall ancestral trees, arrests his attention. The approach to this retreat, known as Craig House, and now used as a receptacle for the highest class of paying patients, is through a gradually ascending avenue bordered with sycamores, the rugged arms of which form a lofty, interlacing arch. The picturesque appearance of the building is heightened by its ancient crow-stepped gables and lofty chimney stacks, its recessed portal, and the green ivy which is seen climbing to the topmost peak of the left wing. From the lawn the prospect is delightful, embracing Edinburgh, the Firth of Forth, and surrounding country, while Arthur's Seat looms up grandly in the distance. The place is historic. On the Borough Muir below, a memorial stone marks the spot where the Scottish standard was unfurled under James IV. prior to the disastrous battle of Flodden Field. It is asserted that Scott took the hill on which Craig House stands as the point of observation for the following description in Marmion :

> "Still on the spot Lord Marmion stayed,
> For fairer scene he ne'er surveyed.
> When sated with the martial show
> That peopled all the plain below,
> The wandering eye could o'er it go,
> And mark the distant city glow
> With gloomy splendor red ;
> For on the smoke-wreaths, huge and slow,
> That round her sable turrets flow,
> The morning beams were shed,
> And tinged them with a lustre proud,
> Like that which streaks a thunder cloud."

This mansion, some three hundred years ago, was the seat of the Carmichaels, a family of distinction in early Scottish annals. Here was the home of Lady Carmichael, one of the four Maries of the ill-fated Mary Queen of Scots, and it is asserted that the unfortunate queen herself frequently graced

by her presence the halls of this venerable edifice. It was
a favorite hunting-seat of James V., and Cromwell, when in
Scotland with his Ironsides, made the place his headquarters. Craig House is in keeping with the troublous times in
which it originated, having been designed for shelter and
defence. It is strongly built, with deep foundations, and
some of the walls, as seen in the vaulted kitchen, are
of great thickness. Interest in the place is enhanced by the
discovery, in later days, of a secret corridor or passage-way
connecting its subterranean vaults with the fields below.

It were difficult to describe the interior of this old-time
mansion with its rambling halls, hidden nooks, unexpected
stairways, and numerous rooms, including spacious saloons
that still retain evidences of their former grandeur. It is
much to the credit of the management that the restorations
necessary to its present use are in conformity with the
period in which it was built. The furnishing and accommodations are on a scale of the highest comfort, if not luxury,
and quite in harmony with the external surroundings. Ascending a flight of richly carpeted stone stairs, one notices
that the carved oak doors are strengthened by iron studs,
and that they have old-style wrought iron hinges and hanging iron handles. The apartment now used as a diningroom is an interesting part of the mansion. The great
arched fire-place, nine feet wide, with its ancient fire-dogs
and bright tile, suggests the generous hospitality of by-gone
days. The dark oak-panelled walls, carved oak mantels and
furniture, and other appointments are in the style of Holyrood Palace. Among other decorations, there is painted on
a stone set in the wall a landscape, in which appears a
castle of mediæval time, with its moat and bridge. In the
lobby adjoining are quaint old paintings, set in the skilfully
carved oak panels of the walls.

The drawing-rooms are on a corresponding scale of elaborate furnishing and decoration. These apartments command a fine view of the northeast, and communicate with a beautiful garden, which is kept in high condition by the aid of patients drafted from the West House. This is a most delightful spot. The grounds are terraced and double banked, and laid out in the fashion of a past period. They were exquisitely kept, and planted with shrubs, creepers, and rare flowers, whose delicate fragrance filled the atmosphere with sweetness.

Violent patients are not received here, and there are no locks and keys, nor bars and bolts, nor other restrictions on personal freedom, except the watchfulness of numerous attendants. It should be stated that the sashes of the upper windows are so adjusted as to permit of being moved only a few inches at the top and bottom. There is an entire absence of gratings or inside shutters. It has been found that patients are greatly benefited, when recovering, by removing them from the surroundings unpleasantly associated with the acute stage of their disease to this retired and quiet place. Besides, the lawn and woods of Craig House afford opportunity for patients from other portions of the asylum to take air and exercise, numbers of whom enjoy the shade and soothing repose of this elevated sylvan retreat. Respecting the desirability of change in surroundings, Dr. Clouston says: "Patients laboring under different kinds of mental disorder, even in different phases of the same attack, are better treated, and have a greater chance of recovery, through having this variety of accommodation. In the acute stage, a case needs more of a hospital; in the convalescent stage, more of an ordinary, cheerful house."

As I left Craig House in the soft repose of a dreamy summer afternoon, the sunlight breaking through the surrounding foliage, as the spirit of benevolence in this institution breaks through the shadows of the unfortunate lives that find a home and refuge here, the fitness of the place for the special purpose to which it is devoted was impressed upon my mind; and with a parting look at "the mansion with a strange history," and the grand old trees about it, I bespoke a blessing on those whose wise forethought had turned this interesting relic of the past to its present beneficent use.

CHAPTER IV.

IRELAND.

THE first asylum for the insane in Ireland was founded by Dean Swift, and was opened in 1745. Although nearly a century and a half has elapsed since its establishment, St. Patrick's Hospital, of Dublin, still continues its work of mercy, the expense of maintaining it being principally met by the revenue derived from its original endowment. The good example set by the Reverend Dean did not, however, stimulate others to the performance of similar acts of generosity, nor did it arouse the Government from its indifference to the needs of the insane. As might be expected in a country for a long time in an unsettled condition and suffering under financial depression, the mentally afflicted were sadly neglected. Illustrative of the misery of those who were restrained in their own homes, a member of the House of Commons, in testifying before a committee in 1817, said:

"There is nothing so shocking as madness in the cabin of the Irish peasant, where the man is out laboring in the fields for his bread, and the care of the woman of the house is scarcely sufficient for attendance on the children. When a strong man or woman gets the complaint, the only way they have to manage is by making a hole in the floor of the cabin, not high enough for the person to stand up in, with a crib over it to prevent his getting up. The hole is about five feet deep, and they give this wretched being his food there, and there he generally dies."

TAKING LUNATICS TO DUBLIN IN THE EARLY PART OF THE NINETEENTH CENTURY.

Subsequently another witness before a later parliamentary committee affirmed that he, too, had 'seen cases analogous to the above.

It is not to be assumed, however, that such practice prevailed generally throughout Ireland ; and where it did exist, it should not be regarded as evidence of intentional cruelty. It must be borne in mind that there was almost an entire lack of asylum accommodation ; that a popular prejudice existed against what were called "mad-houses"; and that the strong love of home and kindred among the Irish people resists the separation of families.

But no excuse can be found for the manner in which the insane poor under public care were treated in Ireland in the early part of the present century. This is as justly subject to censure as was the treatment of the same class in England or Scotland. It appears that it was once the mode in conveying lunatics to Dublin to tie them to the back of a car and compel them to go the entire distance on foot, however long or short it might be. This practice is referred to in the report of a parliamentary commission wherein it is recorded that a physician of note said in the year 1808 : "I give you my honor that of the insane persons sent up to Dublin, almost one in five loses an arm from the tightness of the ligature producing mortification, which renders amputation necessary." Sane and insane were crowded together in the houses of industry or correction. Sometimes as many as three were chained in one bed, and in some instances a chained lunatic and a sane pauper shared a single bed between them. With few exceptions, wherever they were confined there was a lamentable absence of almost every requisite for the peculiar wants of the insane.

In 1804, the attention of the Government was directed to the urgent need of asylum accommodation for the insane poor in Ireland, but nothing was done in the way of extending public relief.

Again, in 1810 the Government made an inquiry into the wants of the insane in that country, the result of which was an appropriation by Parliament for the building of an asylum at Dublin for the reception of insane persons from the whole of Ireland. This institution, opened in 1815, was subsequently enlarged, and is now known as the Richmond District Lunatic Asylum.

Before 1810 there was no regularly organized asylum in Ireland, except that founded by Dean Swift. There were four houses of industry or correction into which the insane were received. These were located at Dublin, Cork, Limerick, and Waterford, but they could not properly be regarded as asylums.

The reports following successive parliamentary inquiries at length led the Government to take energetic steps towards ameliorating the unhappy condition of the insane in Ireland, and the important Act 1 and 2 George IV. was passed in 1821. Lunatics at that time were not provided for except by laws that were measures of police and public safety. In the report of a governmental commission the legal status of the mentally diseased was set forth in the following language: " Lunacy was treated as a crime, and the lunatic as though he was a malefactor, not merely supposed to have committed a crime, but assumed to have been convicted of it." The result of all legislation before 1821 was characterized by a committee of the House of Lords in 1843 as a "distressing example of human suffering, mistaken legislation, and of objectionable practice." Large numbers of the insane, for lack of other accommodation, were confined in the various jails of the country.

The Act of 1821 empowered the Lord-Lieutenant to form

districts, to direct the building of a sufficient number of asylums, and when built to appoint governors of these asylums. The measure further provided for the establishment of a Board of Control who were vested with extensive powers respecting the building and management of asylums. They were finally, with the concurrence of the Privy Council, empowered to frame rules for conducting such institutions. With some alterations and amendments this Act forms the existing system of Irish lunacy administration.

There were erected, under the Act of 1821, nine new asylums, for which the sum of £210,000 was advanced by Government, and which was to be repaid by instalments extending over a series of years, the money bearing no interest. The total cost of the ten district lunatic asylums which had been built before 1835, amounted to £332,207. These provided for 1,837 beds, at upwards of $876 per bed.

From 1843 to 1857 the number of lunatic asylums increased from ten to sixteen. The six new buildings, accommodating 1,760 patients, were erected at a cost of £261,995, or about $720 per bed. The largest of these, that of Cork, was built to accommodate 500 patients, at a cost of about $773 per bed. A Royal Commission, appointed in 1857, reported that the district asylums were to a great extent deprived of their utility as curative hospitals for the insane, on account of the large number of chronic cases which they contained, and recommended that other and separate provision be made for the latter class.

From 1866 to 1869, six additional asylums were erected at a total cost of £292,155, giving accommodation for 1,769 patients. This outlay, as will be seen, was at the rate of nearly $800 per bed, but it was not made in compliance with the recommendation of the Commission, the new asylums having been expensively built as curative establishments, thus leaving the country as before without any special or separate provision for patients of the probably incurable or chronic class. A Commission of Inquiry in 1879, after full investigation, recommended, *inter alia*, that a certain number of the district asylums should be set apart for curative treatment, and that the other district asylums should receive chronic cases requiring special care. The Commission further recommended that the inspection of "lunatics at large," instead of being left to the constabulary, should be undertaken by dispensary district medical officers, acting under the lunacy authorities. The fact of upwards of 6,000 lunatics living "at large" among friends and relatives throughout the country, with no other official oversight than that of constables who collected statistics respecting them and who were guided, in a great measure, by the gossip of the place, was deemed by the Commission to be a serious blot on the lunacy system.

The estimated population of Ireland on the 29th of September, 1886, was 4,889,430. The number of lunatics including idiots under supervision of the Inspectors of Lunacy January 1, 1887, was 14,702, being one to every 333 of the population. According to the last census (1881), the population was 5,159,839. The number of lunatics under the supervision of the Inspectors at this time was 13,444, being one to every 384 of the population. It will be observed that these figures show a considerable decrease in population and an increase in the number of lunatics.[1]

The distribution of lunatics under the supervision of the Inspectors of Lunacy on the 1st of January, 1887, was as follows: In the twenty-two district or public asylums there

[1] The statistics given here as well as those in the chapters on England and Scotland, include only such lunatics as come under the cognizance of the lunacy commissioners and inspectors. This fact should be kept in mind in considering the ratio of lunatics to population.

were 10,077, of whom only a few were paying patients; in the poorhouses (workhouses) scattered over the country there were 3,841; in private asylums 611; in the criminal lunatic asylum at Dundrum 172; and in jail 1. It was estimated that 7,779 of the insane in district asylums were of the incurable class and that about half of the lunatics in the poorhouses were idiots, epileptics, and confirmed imbeciles.

The average weekly per-capita cost of maintenance in district asylums during the year ending January 1, 1887, was 8s. 1d., of which four shillings was paid by the Government. The total amount of the Government grant for that year was £99,608, 16s. In cost of maintenance are included salaries and wages, provisions, clothing, bedding, furniture, household supplies, medicines, incidental expenses, " superannuations," and repairs and alterations. In poorhouses where not more than a bare existence is maintained, the Inspectors estimate that the weekly per-capita cost of supporting the lunatic poor may be set down at 4s. 5½d.

The provision for chronic cases in Ireland is still inadequate, and hence we find more than one fourth of the registered pauper insane restricted to the unsatisfactory accommodations of the poorhouses. The Inspectors of Lunacy, though strongly opposing the common care of insane and other dependents, advise, as an expedient to meet existing emergencies, that in each asylum district a poorhouse be selected in a central position, with land attached, to which tranquil and utterly hopeless cases could be removed and separately cared for. Lunatics in poorhouses are only in a minor degree under the control of the Inspectors, the poorhouses coming under the authority of the Local Government Board for Ireland, which is the central poor-law authority. The Inspectors of Lunacy cannot alter the poorhouse dietary, nor can they order that paid attendants be provided for the insane, though they may direct the removal of a pauper lunatic from the poorhouse to an asylum, should they deem such removal necessary. The Local Government Board has always opposed the conversion, on any large scale, of poorhouses into receptacles for the insane, on the ground that this was never contemplated by the poor-law, which was framed only with reference to the relief of the sane poor.

From an examination made of a considerable number of poorhouses in different parts of Ireland, the writer was convinced that the provision therein for the insane was in almost every respect unsuitable. The patients occupied cramped spaces with imperfect ventilation and without the usual asylum accessories, and there was little or no opportunity for employment or recreation.

There are twenty-three private asylums and licensed establishments. Some of these are charitable institutions supported wholly or in part by voluntary contributions, and others are kept by their proprietors for profit. Among those specially worthy of note are St. Patrick's Hospital, founded by Dean Swift; the Retreat, directed by the Society of Friends; the Stewart Institution; the Hospice of St. Vincent de Paul for females, which is under the immediate charge of a religious sisterhood; and two establishments for men, which are conducted by religious fraternities—one French and the other Belgian.

The Board of Control, created by the Act 1 and 2 George IV., has its office in Dublin Castle. It consists of five commissioners, three of whom are commissioners of Public Works. Two, who are physicians, are the Inspectors of lunatic asylums. The five are joint trustees of the property of district asylums. The powers of the Board pertain more immediately to the purchasing of land, the erection and establishment generally of district asylums, and the furnishing of them. The outlay made under the sanction of this department since 1830 has been about $7,250,000. Subordinate to the commissioners are a secretary, an architect, accountant, and a solicitor. The two Inspectors, with their assisting clerks, have also an office in Dublin Castle. Upon the Inspectors practically devolves the entire work of visitation and supervision of all the lunatic asylums in Ireland. They visit each of the public and private asylums four times a year, besides making special visits to them when necessary. Their inspections also extend to the one hundred and sixty-two poorhouses in different parts of the country. The Inspectors make an annual report to the Lord-Lieutenant, for presentation to Parliament.

The boards of governors of district asylums appointed by the Lord-Lieutenant from among prominent citizens, are similar in many respects to the visiting committees appointed by justices in England, and like them serve gratuitously. The boards of governors supervise the financial affairs of the asylums to which they are appointed, and hold monthly meetings for this purpose.

Among the points wherein the Irish lunacy system differs from the English, the following may be noted: The resident medical superintendent of an asylum in Ireland is appointed by the Lord-Lieutenant, and discharges the duties of his office in conformity with Privy Council rules. There is also a visiting physician attached to each asylum, who is appointed by the Board of Governors. He visits daily the convalescent cases, and signs jointly with the resident physician certificates of discharge. An anomaly is found in magistrates possessing the power to license private asylums or licensed houses, but having no authority to suspend the licenses, the latter power being vested in the Inspectors ; while the Inspectors, not having any voice in the granting of licenses, rarely meddle with what has been done by the magistrates. Then, with regard to pauper lunatics, the relieving officer is not called upon, as in England, to report to the medical officer of his district the existence therein of lunatics or idiots unless they become a charge on the rates. No law of settlement prevails, and the place of commitment must bear the cost of maintenance. In the event of his being convicted of some trivial offence, a pauper or other lunatic may, under a recent statute, be confined as a dangerous lunatic.

While the law is similar to that of England as regards chancery and private lunatics possessed of property, there is a considerable difference as regards pauper or even paying patients of the middle class seeking admission into public or district asylums. As already indicated, nearly all of the patients in the twenty-two district asylums are paupers. One of the regulations made by the Privy Council was that no paying patient should be admitted into any district lunatic asylum so long as there should be unsatisfied the application of any pauper lunatic.

Most of the patients in public asylums are admitted in one of three ways, as follows: First, by application of the friends of a lunatic, addressed to the local board of governors, and accompanied by a medical certificate, and a certificate of identification from a clergyman or magistrate stating that the person to be committed has not means sufficient to support him in a private asylum; second, in cases of urgency the resident physician of an asylum may satisfy himself of the condition of the patient and admit on the bare certificate of a medical man, subject to the approval of the governors at their next meeting; third, by the signed order of two magistrates, which is accompanied by a medical certificate. In the admission of patients to private asylums, two medical certificates and the signed order of a relative or friend are necessary.

The procedure of commitment and manner of conveying patients to asylums are open to grave objections, especially in the case of so-called " dangerous lunatics" committed by magistrates. During the year ending January 1, 1887, sixty-six per cent of the 2,746 insane received into district asylums were admitted in this manner. As bearing upon this subject, the Inspectors say:

" Owing to the hasty procedure too frequently adopted, scanty information, occasionally none whatever, is supplied on the face of the warrants, beyond the offence committed, or the assumption of an intention by an irresponsible agent, such as the breaking of a pane of glass, jumping into a pool of water, or threatening to do violence. This mode of peopling an asylum, however benevolently intended, is upheld by magistrates at Petty Sessions, who frequently have no evidence before them beyond the declaration of parties anxious to get rid of troublesome relatives, and the opinion of physicians casually called in by them to cases with which they may have been unacquainted. Thus not only no reliable information is supplied for the guidance of an asylum physician, but utter strangers are occasionally made chargeable to districts with which they have had no previous connection. Independent, too, of an unfair local taxation, it often occurs that 'dangerous lunatics,' male and female, young and old alike, have to be conveyed from distances extending to seventy or eighty miles, some, mayhap, handcuffed, under a police escort, to their destination. The statute further leads to abuse from magisterial oversight; scarcely a week passes over without illegal committals, and the consequent necessity of our returning them to the committing justices for rectification."

The discharge of patients from district asylums rests with the board of governors, who usually act upon the recommendation of the superintendent. The Inspectors are also vested with the power to order discharge.

Connected with Irish asylums are but small areas of land

for tillage and pasture, consequently there is a lack of some farm products which are nutritious and highly beneficial to the insane. Especially is this true of milk, large quantities of which are required in institutions of this kind. Besides, from the insufficiency of land there is not as much outdoor employment as is desirable. Within doors, in addition to the usual trades carried on by men, such as shoemaking, tailoring, carpentering, etc., basket-making has been undertaken in a few asylums.

Notwithstanding the defects in the Irish lunacy laws, it must be conceded that the administration of the affairs of the existing institutions are generally deserving of commendation. The superintendents are efficient, and manifest an earnest interest which extends to every detail of their work. A few extracts from my notes of visitation will suffice to indicate the general character of the insane asylums of Ireland.

CORK DISTRICT LUNATIC ASYLUM.

This institution is so situated as to overlook a broad expanse of the fertile valley of the Lee, and is surrounded by well-kept grounds ornamented with flowers and shrubbery. The asylum is designed for the insane poor of the district of Cork, and has accommodation for nine hundred and thirty patients. It is built in a semi-Gothic style on the congregate plan. The four-story central department has three-story wings to the right and left, the whole presenting an extended front broken by gables.

The financial affairs of the institution are directed by a Board of Governors composed of forty members. They hold monthly meetings, the average attendance at which is about ten. The asylum is under the immediate charge of a resident medical superintendent aided by two medical assistants. There are also a regularly appointed visiting and consulting physician, an oculist, and a dentist. Included in the house staff are a clerk and his assistant, a matron, house-steward, storekeeper, band-master, librarian, and a fire-brigade instructor.

Aside from single rooms there are various associated dormitories accommodating from four to twenty patients each. A preference was shown for the latter, about two thirds of the inmates being provided for in this way. The bedsteads are made of gas-pipe. Strong canvas, with eyelet-holes around the margins, is laced to the frame-work, and forms a bottom to the bed. This arrangement, facilitating easy removal and cleansing, and thereby affording precautions against vermin, was claimed to have other advantages. Through the centre of some of the large dormitories is a partition with short cross-sections forming two rows of stalls for beds. The partitions are four and a half feet high, and have an open space beneath. The windows have light iron sashes, so adjusted that the upper and lower parts can be simultaneously moved to the extent of four inches. On some of the floors was laid bright-bordered oil-cloth or linoleum. Large open fire-places were mainly relied upon for warmth and ventilation.

The men and women have their meals together in a spacious dining-hall designed to accommodate seven hundred persons. This is also used as an amusement hall, and it has a gallery for the band.

From the several airing-courts at the rear of the building one has a view of the asylum and grounds and of the beautiful valley beyond. The walls of the court for men are built higher at one of the corners for ball-playing, which is here a favorite pastime. The yards for women were made attractive by beds of flowers, plots of greensward, and gravelled walks.

The land connected with the asylum is limited to fifty-seven acres, only thirty-two of which are used for farm and garden purposes, the remainder being taken up by buildings, yards, and groves. Seventy-four of the men were engaged in the cultivation of the ground, mostly by the use of the spade. During the year preceding my visit there were, on an average, forty-six per cent of the patients usefully employed.

It was said that very little restraint was resorted to. Four rooms were padded, both walls and flooring, and in these very excitable patients were occasionally placed. The interior construction of these rooms is designed to obviate the possibility of self-injury. Cribs were not in use. There were, however, box-beds in which patients were sometimes tied. The bathing arrangements include a Turkish bath.

Connected with the institution are two chaplains—one of the Roman Catholic Church, and the other of the Church of Ireland. During the early part of 1886, a chapel for the Protestant inmates was completed and opened for religious services. It is a Gothic structure having colored glass windows, and a tower with a bell. The grounds about the edifice have been graded, planted, and otherwise improved by the aid of the inmates.

The number of patients January 1, 1887, was 476 males, and 449 females.

Brief as it was, my visit to this institution proved an agreeable introduction to Irish asylums. As I left the place, the sight presented was particularly pleasing. The patients were out on the broad and elevated terrace, evidently enjoying the fine music of a band composed of asylum attendants.

BELFAST DISTRICT LUNATIC ASYLUM.

The inmates of this asylum, as in most similar institutions in Ireland, are principally of the dependent class. Here, as elsewhere, the asylum, though primarily designed as a curative hospital, was overcrowded and heavily burdened with chronic cases.

The buildings are plain, substantial brick structures two and three stories high, and have pleasant grounds in front. The asylum estate, containing fifty-five and a half acres, is inclosed by a wall. The buildings, courts, and recreation grounds take up thirteen and a half acres.

On the male side are sixty single rooms. Of the associated dormitories, ten accommodate four patients each ; and eight, sixteen patients each. The arrangements for women are substantially the same as those for men. The bedsteads are of iron with strap bottoms, over each of which is placed a thick straw mat that protects the mattress. At the foot of each bed was a semi-circular willow basket for the clothes of the patient. Special provision is made for epileptics. A night-watch perambulates the building, and makes hourly visits to the patients.

In a large dining-hall, men and women sat at separate tables. There were guards to the open fire-places.

The employments for men include garden and other outdoor work, shoemaking, carpentering, painting, and tailoring ; for women, needlework, knitting, laundry-work, and housecleaning.

The airing-courts are spacious. On the men's side accommodation for playing ball is provided on the plan already referred to at Cork. At their amusements, and in their several occupations, the insane are under watchful supervision. Here, as at Cork, a band composed mainly of attendants played at certain hours, for the entertainment of the inmates.

DONEGAL DISTRICT LUNATIC ASYLUM—LETTERKENNY.

The above asylum, principally occupied by insane poor, is pleasantly situated in the wild and picturesque county of

Donegal. At the time of my visit, it contained 298 inmates under the care of Dr. Joseph Petit, aided by thirty-two subordinate officers, attendants, and servants. Since then the number of patients has slightly increased, there being 324 inmates on the 1st of January, 1887. Only 110 of these were women.

Passing the gateway lodge, which is kept by a patient, one reaches a block of plain yet commodious buildings fronted by a broad green lawn, and commanding an extensive view. The buildings, though not imposing, are pleasingly varied in outline, and are constructed so as to afford the inmates the maximum of sunlight.

There are eighty-six single rooms, six rooms accommodating three each, and ten associated dormitories containing each from ten to thirty beds. The floors were covered with manilla matting. Willow baskets for patients' clothes hung at the foot of the bedsteads. The windows had iron sashes only in the single rooms. The Superintendent favored windows constructed so as to enable the patient when seated to obtain outside views. Open fire-places were noticed in all the principal rooms of the building. The ventilation appeared to be good.

It was stated that restraint is reduced to the minimum in this institution. There were no patients in seclusion at the date of my visit. From the airing-courts, a fine view of the adjacent country is obtained. Dr. Petit advocated the total abolition of walls, which he considered not only unnecessary if there was proper supervision, but positively injurious to the patient. He further maintained that freedom had a salutary influence on attendants, stimulating them to greater attention to duty.

In the way of industries, many of the men were employed on the farm, others at tailoring, shoemaking, etc. The women were, for the most part, engaged in sewing and housework.

It was evident, from an inspection of the whole institution, that it was under good administration.

RICHMOND DISTRICT LUNATIC ASYLUM—DUBLIN.

This is the oldest of the Irish public asylums, and in its management ranks among the foremost institutions of its kind in either Ireland or Great Britain. For thirty years it was under the superintendence of Dr. Joseph Lalor, to whom it is greatly indebted for its present advanced position. At the time of my visit Dr. Lalor was assisted by two medical officers. There were also a consulting and visiting physician and a surgeon. The attendants on the men's side numbered forty-six and on the women's side fifty-four. The patients, of whom there was a slight excess of women over men, numbered about one thousand. There were only a very few paying patients, who were received at a low charge for maintenance. Among the inmates was a considerable number of feeble-minded, on whose educational training much attention was bestowed. The asylum, situated in the midst of well-arranged grounds, is an old-fashioned structure on the congregate plan.

On entering the spacious dining-hall, used also as a place of entertainment, one sees conspicuously placed upon the wall the good old Celtic motto—" Cead Mile Failte " (a hundred thousand welcomes.) There were here many evergreen wreaths left over from the holidays, which lent quite a festive appearance to the interior. Before eating, grace was sung to an organ accompaniment. The table furnishing consisted of an ample supply of crockery, knives with blunt points, and forks with short prongs. The food was wholesome, but there was little variety.

The associated dormitories are large and airy, while many of the single rooms were each found to afford 1,200 cubic feet of air space. The mattresses were uncovered and the bedding folded and laid upon them for convenient inspection. The ordinary sheeting was of strong Irish linen. Epileptic patients sleep in deep box bedsteads. Two nurses are on duty at night in each of the large dormitories, and two are specially charged with the oversight of epileptics. The night watch is checked by tell-tale clocks. The rooms for seclusion were padded with cocoa fibre, and the edges of the floor matting were bound with leather. In many of the day-rooms, a strip of wood was fastened round the margin of the floor to prevent the furniture from being pushed against the wall. *Papier-maché* basins were provided for violent patients.

In the day-rooms for women there were many flowers in hanging-baskets before the windows. The rooms were cheerful, and comfortably furnished. From the furniture all angular corners were rounded off to prevent injury to the patients. The chairs were of strong hard wood. Birds in cages were noticed in all the day-rooms of the institution. In the departments for hysterical patients and those suffering from melancholia, there were numerous pictures and many other objects varying in form and color, to attract and interest the inmates. The windows are so constructed as to open simultaneously at the top and bottom. Some of the windows have iron sashes. In all the wards were seen open fire-places.

Respecting the efficacy of employment and training as aids to the recovery of persons suffering from mental disorder, Dr. Lalor said : " Employment of some kind is the agent to which we should look for the improvement and cure of the insane." He has been designated " the father of the school system as applied to asylums." In this connection it should be mentioned that, in addition to other employees, there are two male and three female school teachers, whose duty it is to instruct patients in the ordinary rudimentary branches and in music and drawing. There are besides ten "school attendants" who are employed as ordinary attendants, but who, during school hours, assist in school work. For the performance of these duties such persons are selected as have had experience as teachers and who, as Dr. Lalor expresses it, have thus acquired the habit of exercising a mastery over the mind, and a practical knowledge of regulating the habits and conduct which teaching confers— advantages which not only come in play during school hours, but in the general management of the institution. The main objects kept in view here are to provide varied occupations for as many as are able to work; to apply a system of education that will divert and strengthen the mind; and to promote, by every conceivable means, the happiness and welfare of the inmates. During the year preceding my visit no fewer than 470 men and 467 women patients had attended the several classes. The school is a kind of kindergarten. There are lessons on color, subjects in natural history are illustrated by pictures or real objects, and, in general, material things are used to develop ideas. The school is conducted in three divisions for each sex, and these are subdivided into six classes for males and five for females. Fifteen minutes at the opening are given to an "inspection as to cleanliness." The lessons are brief, and there are frequent physical exercises and singing, with intervals of outdoor and indoor recreation. The course of instruction is varied so as to constantly occupy the mind and form an agreeable routine for the whole day, and the exercises of different days are unlike. The following table shows to what extent the school system obtains in this asylum, and partly indicates how the time of the patients is occupied during a year:

Occupation of Time.	Average number of patients.			Highest number during the year.		
	M.	F.	T.	M.	F.	T.
Reading...............................	226	280	506	258	290	548
Writing on Slates.....................	40	200	240	53	220	273
" on Copies, from Books or Head-Lines......................	100	80	180	120	86	206
" from Dictation................	175	120	295	180	130	310
Grammar, Parts of Speech............	118	150	268	122	156	278
" Parsing and Derivations......	30	24	54	32	30	62
Geography, Local and Descriptive......	230	130	360	236	140	376
" Physical and Historical.....	30	36	66	32	38	70
Natural History, with Illustrations......	250	130	380	256	136	392
Mechanical Powers, from Models.......	55	—	55	59	—	59
Arithmetic, Notation and Simple Rules..	139	146	285	145	150	295
" from Ball-Frames..........	43	96	139	46	100	146
" Compound Rules...........	142	36	178	147	40	187
" Proportion and Interest.....	30	2	32	32	4	36
" Mental...................	226	60	286	258	66	324
Drawing Class........................	40	2	42	45	2	47
Object Lessons.......................	226	120	346	258	130	388
Marching to Music....................	190	280	470	200	290	490
Physical Exercises and Drill...........	190	20	210	200	24	224
Mixed Concert (bi-weekly)............	36	70	106	40	74	114
Concerts (fortnightly).................	36	30	66	40	33	73
Attending Singing Classes daily........	54	100	154	60	130	190
Assist in Teaching....................	12	12	24	14	14	28
Country Walks (weekly)..............	90	80	170	100	108	208
Visiting Zoölogical and Botanical Gardens.	80	35	115	80	42	122
Assist in Keeping Accounts, Record of Occupation Tables, and Copying........	6	1	7	8	1	9
Attending Review on Queen's Birthday..	132	—	132	—	—	—
Outdoor Recreation (daily)............	293	280	573	300	290	590
Attending Religious Instruction in their Respective Churches. { R.C. 80 141 / I.C. 18 29 }	98	170	268	120	200	320

In connection with the asylum are but fifty-four acres of land, fourteen of which are taken up with buildings, yards, etc. The farm for the men and the laundry for the women are valuable outlets for expending the superfluous energy of the patients. Most of the washing is done by hand, and the garments are dried in the open air. The clothing of the filthy patients is, however, washed by machinery in a special department. After soaking for six hours in large vats, it is subjected to a high temperature and the action of chemicals. The following returns show the daily average number of patients employed and unemployed for a year:

Male employment.	Number employed.	Female employment.	Number employed.
Attending School only...	143.7	Attending School only...	26.
Office Work............	3.0	Assisting in Stores.......	9.
Garden and Farm Labor.	123.2	Assisting in Laundry.....	48.
Shoemaking............	7.5	Needlework.............	130.
Carpentering...........	7.1	Darning and Mending...	135.
Painting...............	9.0	Fancy Work............	8.
Tailoring...............	7.6	Learning to Sew, Knit, and Crochet..........	14.
Mattress Making........	9.0		
Mason-Work...........	4.4	Quilting & Machine Work.	2.
Smith-Work and Engine.	3.0	Assisting in Kitchen and Dining Hall..........	18.
Plumbing..............	1.4		
Sweeping Chimneys.....	1.0	Cleaning House.........	69.
Patients' Dining Hall and Food Van............	17.0	Attending Members of Staff..................	20.
Cleaning House.........	55.6		
Attending Members of Staff................	6.2	Total Employed Daily.	479.
		Unemployed..........	80.
Total Employed Daily.	398.7	Total.........	559.
Unemployed..........	54.3		
Total........	453.0		

Amusements, books, and periodicals are provided in almost endless variety. The following list indicates some of the kinds of recreation : Billiards, bagatelle, backgammon, cards, chess, dominoes, draughts, Aunt Sally, happy families, pumble-chook, counties of England, conjuring, music, singing, archery, racing games, snap, cricket, football, lawn-tennis, skittles, nine-pins, walking in the garden, walks to Phœnix Park, to the Zoölogical and Botanical Gardens, and fortnightly parties. The number habitually taking part in amusements, games, and reading during the year previous to my visit was 400 male and 490 female patients. In one of the wards, about thirty patients were seated in a circle round an organ, listening with rapt attention to the music. Women were sewing under a trained teacher in another room. Some displayed not a little skill in making artificial flowers. In all of the departments of this institution there appeared an unusual air of activity without disturbance. There was constantly during the day some kind of amusement, employment, instruction, recreation, or entertainment. The whole system seemed to be ingeniously planned to attract the attention of the patients and turn their thoughts from themselves, and at the same time to inculcate habits of self-control.

It being visitors' day, friends were seen holding free converse with the patients. There is a rule that no patient shall be seen by any visitor until after a month's residence in the asylum.

In this large institution, there were, on the day of my visit, only one man and one woman in seclusion. It was said that the refractory are confined but for a short time. Dr. Lalor expressed his opinion strongly in favor of the efficacy of sunlight for secluded cases. Patients liable to kick others wear soft shoes. Violent and disturbed women were, for the most part, plainly dressed in linsey-woolsey. In the refractory division some of the patients were picking cocoa fibre; others, stuffing and tufting mattresses. In this ward, fire-guards were placed in front of the fire-places. Among the women outside, some were seen wearing a crimson turban-like head-dress. This was so tastefully arranged as to be attractive; nevertheless, it was a device humanely designed to protect epileptic patients from accidental injury.

The bearing of the attendants appeared to be kind and considerate. The salaries of the women ranged from £11 to

£15 per annum, with clothing and food, and those of the men from £18 to £25, beginning at the lower figures and increasing to the maximum.

On the grounds are two churches—one a large and artistic structure for Roman Catholics, the other a tasteful building for those belonging to the Irish Church. For each of these and for the Presbyterian inmates there is a regular chaplain.

Not content with discarding old methods of restraint, Dr. Lalor also signalized his advent in this institution by pulling down all the interior walls of the yards and airing-courts. These are only some indications of the various ways in which the asylum management was greatly improved by this distinguished alienist supported by an enlightened governing Board.

Since my visit here, this eminent philanthropist and true friend of the insane has passed from the field of his useful and successful labors. The asylum is now superintended by Dr. Conolly Norman. On the 1st of January, 1887, it contained eleven hundred and three patients.

CHAPTER V.

CONTINENTAL COUNTRIES.

TURNING our attention to the insane on the Continent of Europe, previous to the reforms introduced in the latter part of the eighteenth century, we find that this wretched class were subjected, if possible, to even greater cruelty and neglect than that under which they then suffered in the British Isles. Harmless lunatics were exposed to severe and lawless chastisement; the troublesome, when held in custody, were confined in unwholesome prisons, where they were sometimes placed under the charge of criminals, who, with whips and savage dogs, held barbarous sway over the helpless objects of their cruelty. Chains were considered indispensable; even the garrote was used; while privation and uncleanness were characteristics of these forlorn places.

Riel, in writing of the German asylums in 1803, said: "They are madhouses, not merely by reason of their inmates, but more especially because they are the very opposite of what they are intended to be. They are neither curative institutions, nor such asylums for the incurable as humanity can tolerate. They are for the most part veritable dens. Has man so little respect for the jewel which makes him man, or so little love for his neighbor who has lost that treasure, that he cannot extend to him the hand of assistance and aid in regaining it? Some of these receptacles are attached to hospitals, others to prisons and houses of correction; but all are deficient in ventilation, in the facilities for recreation; in short, they are wanting in all the physical and moral means necessary to the cure of their patients."

In 1818 Esquirol said of the insane in France: "I have seen them naked, or covered with rags; with nothing but a layer of straw to protect them from the cold dampness of the ground on which they lay. They were kept on food of the coarsest kind; they were deprived of fresh air to breathe, and of water to quench their thirst, and even of the most necessary things of life. I have seen them given up to the brutal supervision of jailors. I have seen them in their narrow cells, filthy and unwholesome, without air or light, chained in such dens as one might dislike to confine ferocious beasts in."

The condition of the insane in the most refined and luxurious of European capitals, just before the changes effected by Pinel, is thus described by Pariset: "In spite of the reforms attempted under the most humane of all kings, the hospitals of the Capital were still in a deplorable state of barbarity. The one which presented the most revolting aspect was the institution of Bicêtre. Vice, crime, misfortune, infirmity, diseases the most disgusting and the most unlike, were there confounded under one common service. The buildings were unfit for human habitation. Men, covered with filth, cowered in cells of stone, narrow, cold, damp, without air or light, and furnished solely with a straw bed that was rarely renewed, and which soon became infectious,—frightful dens where we should scruple to lodge the vilest animals. The insane thrown into these receptacles were at the mercy of their attendants, and these attendants were convicts from prison. The unhappy patients were loaded with chains, and bound like galley slaves. Thus delivered, defenceless, to the wickedness of their guardians, they served as the butts for insulting raillery, or as the subjects of a brutality so much the more blind, as it was the more gratuitous. The injustice of such cruel treatment transported them with indignation; while despair and rage, finishing the work with their troubled reason, forced from them by day and night cries and howlings that rendered yet more frightful the clanking of their irons. Some among them more patient or more crafty than the rest showed themselves insensible to so many outrages; but they concealed their resentment only to gratify it the more fully. They watched narrowly the movements of their tormentors, and surprising them in an embarrassing attitude, they dealt them blows with their chains upon the head or the stomach, and felled them dead at their feet. Thus was there ferocity on the one hand, murder on the other.

"At the Salpêtrière were received only such as had undergone treatment at Hôtel Dieu,—the common and imperfect treatment which rendered the state of the patients more difficult and dangerous. To restrain their fury, they were crushed under the same rigors, or rather they were irritated by the same sorts of violence. Sometimes enchained naked in the almost subterranean cells, worse than dungeons, they had their feet gnawed by rats, or frozen by the winter's cold. Thus injured on all sides, their embittered hearts breathed only vengeance, and intoxicated with frenzy like the bacchantes, they burned to tear in pieces their attendants, or to destroy themselves before them." The reforms effected by Pinel at these two dismal places have already been described.

It is gratifying to every lover of humanity to reflect that over the entire Continent great progress has been made in legislation for the protection of the insane. Humane

BELGIAN CAGE.

methods have displaced cruel treatment, and with few exceptions well-equipped asylums have taken the place of cells and dungeons; the private religious establishments, useful in their day, have been almost entirely superseded by public institutions; while advances made in medical science are shown in the greater effectiveness of the present curative treatment of the mentally diseased. The first asylum in central Europe established exclusively for the insane was opened at Vienna so recently as 1784. Curative treatment as the primary aim of institutions for the insane was soon after recognized in Germany, and improved methods of care for this class were generally adopted. Among the notable reforms of recent years were those accomplished in Belgium. Following an official inquiry, the Government, in 1850, made radical changes in its laws affecting the insane. It virtually reorganized its lunacy system and abolished many grave abuses that had previously existed in that country. Under the new system the use of cribs or cages then common in asylums was forbidden. One of these peculiar contrivances for confining the insane was on exhibition at the National Fair held at Brussels in 1880. It was a sort of wooden cage on short posts, into which the food was passed through a small opening in the frame-work.

It is, however, neither within the aim nor the scope of this work to follow the progress of beneficent reforms affecting the insane, nor will an attempt be made to describe the various lunacy systems of continental countries; but brief reference will be made to certain features of those of France, Belgium, and Prussia.

In France the insane poor and orphan and abandoned children are made objects of special care by the state, which readily accepts the obligation of supporting them when relatives are unable to do so. The long-established lunatic asylum at Charenton is under the immediate direction of the National Government, and is open to patients from all parts of the Republic. With this exception, the public institutions for the insane are local foundations. Each of the eighty-seven political divisions of France called departments must provide a public institution for persons of unsound mind ; or else, with the approval of the Minister of the Interior, make an agreement with a public or private asylum in the same or an adjoining department to receive its insane poor. The department, aided by contributions from the commune in which was the home of the patient, defrays the expense of maintaining its insane paupers when the relatives are unable to meet the charge. The proportion paid by the commune depends upon the amount of its revenue.

The sanction of the government is necessary to the establishment of a private asylum, and the person applying to the Prefect for a license to open such an institution must be of good moral character. He must have received the degree of doctor of medicine, or must show that he has secured the services of a properly qualified physician to supervise the medical department of the proposed asylum. This physician must be approved by the Prefect, who may at any time, with the concurrence of the Minister of the Interior, discharge him. Before opening any asylum for the insane it is necessary that the rules respecting its interior management be submitted to the Minister of the Interior.

Public and private asylums are periodically inspected by a committee consisting of the Prefect of the department, the President of the local Tribunal, the local Procureur of the Republic, the Judge of the Peace, and the Mayor of the Commune. Public asylums are under the supervision of a committee of five persons appointed by the Prefects, of whom one retires each year. They hold meetings monthly, or oftener if necessary. It is the duty of the Procureur of the Arrondissement to visit every public asylum in his district once in six months and every private asylum once in three months, or oftener if he thinks it necessary. The asylums are also visited by the Prefect and other officials who are required to report to the Minister of the Interior respecting the condition and treatment of the insane.

Patients are consigned to asylum care in two ways. The first is compulsory, and applies to dangerous lunatics and all persons, who, if at liberty, might endanger the public safety. These are dealt with as a measure of police by the public authorities—in Paris, by the Chief of Police, and in the departments by the Prefects. In very urgent cases the Commissioners of Police or the Mayors are authorized to take provisional measures for restraint, but the matter must at the same time be reported to the Prefect. The second form of admission is called "*placement volontaire.*" By this method a relative or friend applies to the proper authorities for the admission of a lunatic into an asylum. The application must be accompanied by a medical certificate indicating the mental condition of the patient and stating that he is insane and should be confined in an asylum. The physician who gives the certificate must not be a relative of the patient nor connected with the asylum to which he is committed, and the certificate must be signed within fifteen days after the physician has seen the patient.

Every person confined within an insane asylum must be dismissed therefrom whenever the physician of the institution asserts that a cure has been effected. A husband or wife, or more distant connection, or others may demand the removal of a patient from an insane asylum, and the demand must be complied with unless the physician of the establishment thinks that the liberty of the patient would compromise public safety and personal welfare, in which case he makes known his apprehensions to the Mayor, who immediately forbids the dismissal. The provisionary prohibitory order expires in two weeks unless the Prefect, who is informed of the case, gives orders to the contrary. The Prefect may at any time demand the immediate dismissal of a patient placed in an insane asylum by the voluntary application of friends.

A peculiarity of the French and Belgian lunacy laws is the liberal conception of the right of the patient to his liberty—a right under the French code, entitled that of "*Reclamation*," by which a patient himself, or some relative or friend, or any person interested in him, may claim his discharge before a civil court. Of this unique feature of jurisprudence, the late Dr. Achille Foville, Inspector-General of the Department for the Insane in Paris, says:

"Every one placed in a lunatic asylum may claim his discharge before the civil court as often as he pleases, and at any time. Thanks to this right, every patient is at liberty to have his mental condition examined by the law of his country, without delay, without complicated formalities, and without being hindered by pecuniary considerations. . . . This right of claim is evidently the strongest guarantee which the law can give to individual liberty. It is no less valuable, I am convinced, to asylum physicians. In my opinion, it is a very mistaken view which some of them take, who regard it as a mark of personal distrust, and wish it to be resorted to as seldom as possible. I, on the contrary, think that this process furnishes alienist physicians with the best means of minimizing their responsibility of clearing themselves from every imputation of abuse of authority, and of displaying, in all circumstances, the sincerity of their opinions and the rectitude of their conduct."

A marked peculiarity of the Belgian system of caring for the insane is that of boarding them in families under governmental supervision, as in the colony of Gheel. A new colony, Lierneux, similar to Gheel, has recently been begun in the Walloon district, not far from Liége. The number of insane in the asylums of Belgium on the 1st of January, 1886, was 8,986. This country, which is one of the most densely populated in the world, contained at the date named nearly six millions of inhabitants.

The support of pauper lunatics in Belgium must be defrayed by the communal treasury in default of sufficient charitable endowments. If there is proof that the revenues of a commune are inadequate to this purpose, the province is bound to meet the requirement. There are three government asylums having accommodation in all for upwards of 1,800 patients. The municipal asylums of Bruges provide for about 1,300 insane. Besides, there are numerous smaller hospitals kept by individuals or religious communities.

Licenses are required from the government before opening a public or private asylum. One or more medical men must be connected with the institution, and the patients must be visited daily.

A medical certificate signed by a physician not connected with the asylum to which the patient is committed is requisite to admission. The certificate must describe minutely all the peculiar symptoms of the patient's insanity, must state its duration, and give particulars as to previous treatment. A sealed statement accompanies the physician's certificate, declaring the cause of insanity and whether any other member of the patient's family has been similarly afflicted. The medical certificate may be dispensed with in urgent cases, but one must be obtained within twenty-four hours after the patient is received into the asylum. Certificates for the admission of pauper patients are made by the medical officer of the poor.

Notice of a patient's reception into an asylum must be forwarded, not later than one day after reception, to the Provincial Governor, the Attorney-General of the Arrondissement, the Cantonal Judge of the Peace, the Burgomaster of the Commune, the Committee of Inspection, and the Secretary of the Permanent Commission. The sixth day after a patient's reception into an asylum the physician transmits his observations from daily examinations of the case to the Attorney-General, and eight days thereafter he must inform the Secretary of the Permanent Commission of his patient's condition. Two medical men, members of the Permanent Commission, are required to visit the patient at intervals of two months during the first half year that he is an inmate of an asylum, the first visit to be made within three days after notice of commitment, and a copy of the results of their examinations must be sent to the Secretary of the Commission. The second half year but one visit has to be made by these officers. Persons are appointed by the Government to visit all asylums as follows: Every three months they are inspected by the Attorney-General of the Arrondissement; every six months by the Burgomaster of the Commune; every year by the Provincial Governor, or by a member of the Provincial Council nominated by the Governor. In each Arrondissement a special committee, composed of several members, are required to visit annually all asylums coming under their immediate jurisdiction; and three Government Commissioners are obliged to make an annual report from personal observation of all asylums and every thing relating to lunacy in Belgium. Notice of discharge, containing the patient's name, a statement of his mental condition at the time of discharge, the name of the asylum in which he had been confined, and of the house in which he is going to reside, must be sent within twenty-four hours after discharge to the same persons to whom notice of admission was sent.

There are no general laws regulating the care of the insane in Germany, and the manner of providing for them varies in the different states. Throughout the Empire, the institutions for this class are small as compared with those of England, the United States, and some other countries. Many of the buildings used for asylums before 1850 were once monasteries. It is estimated that only about one third of the insane in Germany are given asylum treatment, the remainder being considered suitable for family care. The number of public institutions for the insane in the German Empire in 1886 was 103. Of private asylums there were 69. Besides these there were 29 of what are called open asylums (*Offene Anstalten*), which receive cases of nervous disease and the lighter forms of mental disorder. There were also 32 private establishments for idiots or the feeble-minded.

In Prussia, the insane are mostly committed to asylums according as they are curable or incurable; but whatever their mental state, it must be set forth, together with other facts, in a physician's certificate, which is accompanied by a statement of particulars respecting the family relations of the patient, and bearing the signature of the authorities. Most asylums receive curable patients upon the request of a near relative or other intimate acquaintance, even though such person has no legal authority over the patient, provided the consent of the local authorities is obtained. Incurable patients are placed under legal guardianship, and the petition for their reception into an asylum is made by the guardian or other legal representative. In cases of urgent necessity, however, temporary commitment may be made without such petition. Proprietors of private asylums are required to present to the proper authorities an annual report respecting the condition of their institutions, to keep a daily register according to a prescribed form, and to make a special statement regarding each patient.

The various provinces have different regulations respecting the reception of incurable paupers into their asylums. Some receive such free, or the expense is met by funds from the general revenue of the province; others have a fixed number of free places, and any demand in excess of these must be met by the community to which the pauper belongs; still other provinces will receive incurable paupers only after the expense of maintenance has been guaranteed by the authorities of their places of settlement. Preference is given in the free places to dangerous cases.

A brief description of a few representative institutions is here presented, as illustrating different modes of caring for the insane. Deserving of special attention are the colony system of Clermont en Oise in France, the colony of Gheel in Belgium, and the Alt-Scherbitz asylum in the Province of Saxony, Prussia. The two last are separately dealt with in subsequent chapters.

GAUSTAD INSANE ASYLUM—NORWAY.

In the care of the insane there is little to be gleaned from Norway, if we accept as a criterion the Gaustad asylum near Christiania, the capital of this northern region. An examination of this institution produces in the mind of the stranger unpleasant sensations, and for aught there is of ocular demonstration to the contrary, it might have been built for the retention of some of those stormy characters of Scandinavian mythology, whose wild doings are familiar to us through ancient song and saga. The massive construction of the refractory wards seems sufficient to resist the hammer of Thor himself. Indeed, it is questionable whether the ponderous weapon of the god of thunder could have made any sensible impression on any part of these towers of strength.

The Gaustad asylum, pleasantly situated among the green hills, is reached after a short drive northwesterly from Christiania. This is one of the three public asylums of Norway. It was established about the middle of the present century, since which time those at Rotvold and Eg have been erected. The brick buildings with slate roofs are mainly of two stories. There is an administration building, back of which to the right and left are two opposite ranges of blocks connected by corridors. The women occupy one side of the asylum and the men the other. On the day of my visit the inmates numbered 168 men and 152 women.

The several departments were warmed by stoves. Excepting the provision made for first-class patients, the sleeping arrangements were mostly on the plan of associated dormitories. The floors and walls were generally bare and the rooms presented a cheerless aspect. There appeared, moreover, to be a lack of supervision, and the treatment seemed severe. In one of the halls nearly half of the men were lying on the floor. In one of the sitting-rooms not less than twenty were in that position. There was also a dreary and neglected appearance about the small yards or airing-courts, which are inclosed by high stone walls.

All of the outside windows had iron gratings, except in the refractory rooms, where the windows were high above the floor and out of the reach of the patients. The arrangements for ventilation were faulty. For discipline, an immense douche was provided, the water falling ten or twelve feet through a one-and-a-half-inch pipe. In the refractory ward the massive bedsteads were of two-inch solid oak, as were also the tables and benches, which were firmly secured to the floor by iron fastenings. The doors were solid and strongly barred. There were a number of canvas-padded rooms.

Many of the women wore strong dresses fastened at the back. The garments generally were of substantial material. In several of the yards a considerable number of women were lying on the ground. In the women's workroom some were engaged in spinning, sewing, or knitting. For the use of the better class of quiet patients there are ornamental grounds inclosed by a picket fence about eight feet high.

The state asylums of Norway are controlled by the Minister of Justice, and the local asylums by the communal authorities. Many of the insane are boarded out by poor-law committees, whose proceedings are supervised by a commission. All persons of unsound mind have a legal claim upon the state for relief. Most of the lunacy legislation in Norway, as well as in Sweden and Denmark, has been with reference to committal and discharge, and more attention has been given to the protection of society than to the rights of the insane.

CONRADSBERG INSANE ASYLUM—SWEDEN.

Some of the hard and forbidding aspects of the Gaustad asylum in Norway are noticeable, though in a less marked degree, in Sweden. The Conradsberg asylum, on the outskirts of Stockholm, although a modern institution, presents little that is instructive, whether we view its external surroundings or its internal furnishing. Both acute and chronic cases are here cared for. The buildings are of two stories, with basement, and have lateral wings projecting from the central or administrative department.

This asylum contained, on the day of my visit, 252

patients—140 men and 112 women. All who are able are required to pay, and districts sending patients are responsible for their maintenance. First-class patients are received at two *kroner*,[1] and second class at seventy-five *öre*[1] per day. The latter live in associated wards, as do also the third class, who are paid for by the authorities. The dress of the second and third class was uniform. There were about twenty first-class patients, thirty-five of the second class, and the remainder were third-class patients.

The rooms of the first class had no carpets, and were plainly furnished; but here, as in the second-class department, were settees, also a few pictures on the walls. The bare floors and scant furnishing do not impress one favorably, though it must be acknowledged that a condition of scrupulous cleanliness was everywhere observed. The dormitories of the second and third class contained from three to eight beds, with narrow wooden bedsteads of the French box pattern.

There are two refractory wards, one for each sex. These contained about thirty inmates, with one attendant to every eight patients. The single rooms of the refractory ward, like the larger apartments, have windows so high as to be above the reach of patients. The bedsteads were of the box pattern with sides two inches in thickness, and they were firmly secured to the floor. The doors in this ward were very solid, being three inches thick. Apertures were provided for observation of the occupants of the single rooms in this department. Heating was effected by means of hot-water pipes.

A large, well-lighted, and pleasant apartment was used as the women's workroom. Here were two tables, around which patients were busily sewing. In the way of industries, besides sewing and spinning, something is done at weaving

---

[1] A *krone* (100 *öre*) is equivalent to about twenty-six cents.

linen. The men do a little at carpentering, shoemaking, painting, and general repairs.

There were no padded cells. No mechanical restraint came under my observation. The airing-courts were small and inclosed by high fences.

A contracted recreation yard that I entered had nothing cheerful about it, and resembled those of the insane departments of some of our country poorhouses. There was but little grass in the yard, owing to the constant tread of feet. In a pavilion furnished with benches many patients were seen lying down. In the women's department were some shade trees, and the patients were here provided with wooden benches and settees having backs. In a small, plain-looking chapel, furnished with benches without backs, Sunday services were held.

The patients, classified as to mental condition, dined in association, except in the case of the refractory, to whom the food was conveyed. Wooden spoons were used for the latter class. The ordinary dining-rooms are on the second floor. The tables were of two-inch hard wood. In some cases knives and forks were furnished, in others only spoons were allowed. The knives, where used, were of the ordinary table-knife pattern. The dinner for the second and third class patients consisted of soup made of barley and vegetables, fresh beef and potatoes, rye bread and butter. Their breakfast was rye bread and butter with milk. It was stated that the supper would be oatmeal porridge and milk. Small beer was allowed to be used freely during the day.

This asylum, like all others in Sweden, public and private, is controlled by the government medical authorities.

ST. HANS HOSPITAL.—DENMARK.

A short ride westerly by rail from Copenhagen brings one to this well-known institution, which is reputed to hold the most advanced position of the Danish asylums. Like that of Aarhus and Vordingborg, it receives both acute and chronic patients from allotted districts, while the asylum at Viborg is specially designed for chronic insane of the whole kingdom. Nothing further is necessary to admit a patient into any of these establishments than the certificate of a duly qualified physician. There are no special lunacy laws governing asylums, nor is there any systematic or regular governmental inspection or supervision of the insane. The St. Hans Hospital is supervised by a committee of magistrates, and is under the immediate charge of a resident physician and superintendent, who is assisted by a strong staff of medical officers. At the time of my visit there were in this institution 400 men and 340 women divided into first, second, and third classes.

The buildings are detached, and comprise a number of separate departments. That devoted to administration purposes forms one; the hospital for the treatment of acute cases, another; a third is used for chronic male patients; and a fourth is occupied by chronic female patients; while the kitchens and wash-house are also separate structures. The surrounding grounds, embellished with flowers and shrubbery, include a park and lake. There are also an abundance of shade trees and an extensive system of gravelled walks. The grounds belonging to the hospital department are especially remarkable for their elaborate ornamentation and careful keeping. A portion of the estate, which contains about ninety acres, borders on the Rœskilde Fjord.

The asylum buildings are chiefly of three stories, with day and work rooms below, and sleeping-rooms above. It was observed that the sash most generally used was of iron and had small square panes. The institution has about eighty single rooms; the remainder are associated dormitories, with, in some cases, sixteen beds in a room. The first-class patients occupy single rooms; the second small, and the third larger associated dormitories. The single rooms are of good size and well lighted. For the seclusion of unquiet patients there are about fifty rooms, the windows of which are protected by wooden shutters. The bedsteads were of wood and of the French or box pattern. The bedding was plentiful, and included a mattress laid on straw. The male and female patients have their meals apart in large dining-halls, and so rigid is the separation that two kitchens are provided—one for men and the other for women. Food is conveyed from them to the dining-rooms in wagons constructed for the purpose. There is chapel accommodation, and there are also pleasant amusement rooms, including a billiard-room, and a bowling alley. The sitting or day rooms were in general supplied with books and other reading matter.

Various plans are devised to keep all occupied who are able to work. Among the women were some who objected to every employment except knitting, and their labor in this direction more than met the requirements of the institution. It was therefore directed that stockings should be knit not only for the inmates, but also for sale. During the year preceding my visit eleven hundred pairs were made in excess of the wants of the institution. To further engage the inmates in this direction knit undershirts were substituted for a woven fabric previously purchased for them.

Among the men who are able and willing to work are many mechanics who are employed at their various trades. All the necessary painting, blacksmithing, and carpentering is done by them, as also the tailoring. Their labor is also made use of in constructing sewers and water-works, in draining the land, clearing up underbrush, and making gravelled walks; also in converting the woods belonging to the hospital into pleasure grounds, and carefully tilling and caring for the grounds generally. The bucket-system is in use here, and the asylum waste is utilized to benefit the land. Outdoor work, when the season and weather permit, is substituted for the regular indoor employment of manufacturing wooden shoes and paper boxes; and by this thorough systemization of labor the cost of maintenance is reduced, while the grounds and buildings are beautified and improved.

Beach bathing, which the patients greatly enjoy, is an advantage possessed by this institution. Even those who most oppose any prescribed treatment yield more readily to the strand bathing in summer than to the warm house baths in winter; and the promenade to and from the beach is looked forward to as one of the pleasures which the summer months will bring. On account of limited accommodations at the beach it is impossible to bathe all the patients every day; therefore the baths are divided into sanitary and ablutionary. The chronic cases are bathed but once or twice a week, while those who seem really benefited receive baths daily. The superintendent, Dr. Steenberg, says: "There have been abundant occasions when I could date the beginning of convalescence, or at least essential improvement both in body and mind, from the day when the patient began strand bathing." The water of the Fjord on which St. Hans Hospital is immediately situated is clear, but the sea bottom near the shore is muddy and rough with sharp stones; therefore it is necessary to float the bath-house some distance from the shore. For this reason it is deemed unsafe to allow patients who are likely to commit suicide the use of the shore bath without special precautions being taken. The superintendent says: "Although during all the years I have been here but one person has met his death by drowning, I cannot deny that many a time I am not a little anxious for patients with acute melancholia to whom I permit the use of the shore bath on account of its unquestionable beneficial effect upon the disease." As affording facilities for bathing a larger number of patients, including some of pronounced suicidal tendencies, a bath-yard in the Fjord was under consideration.

The institution appeared to be well organized, and the treatment seemed kind. A large number of attendants were employed, and it was said that restraint was very rarely resorted to. A highly favorable opinion was derived from the brief visit made to this asylum.

FRIEDRICHSBERG ASYLUM FOR THE INSANE, NEAR HAMBURG, GERMANY.

This large and modern institution is under the control and direction of the authorities of Hamburg. At the entrance to the grounds is a snug porter's lodge, and near by is the residence of the chief physician. A little to the right is the "Pensionnat," a large building occupied by boarders of the first and second classes. Farther on to the left is the principal edifice, which is symmetrically and plainly built of red brick, the central structure being three stories and the lateral portions two stories in height. The administrative department occupies a middle position between the male and female sections of this building. In front is a large park used for recreation and promenades. Rearward are cultivated grounds affording outdoor industrial employment for the insane.

At the time of my visit the institution contained 1,010 inmates. These included acute and chronic cases. There are four classes or grades of patients, received at rates varying according to accommodation. Apart from these grades there is a further classification into sections, as follows: 1. Incurables; 2. Those curable and able to work; 3. New arrivals; 4. Refractory and such as require close supervision. The plan of treating curable with incurable insane patients was disapproved of by the officials.

For the four classes paying different rates the arrangements vary. The first class have, according to stipulation, one apartment with mahogany furniture and one attendant to each patient. The number and variety of meals being rather singular, the dietary is given for a single day. The food is slightly changed on three days of the week. The first breakfast for this class consists of an allowance of prepared coffee, cream and sugar, biscuit or rusk. Their second breakfast comprises an allowance of soup or tea, butter and extras, including meat, white bread, cheese, etc. At noon they partake of soup, meat or fish, fowl, game, potatoes and other vegetables. In the afternoon they sit down to a meal resembling the first breakfast. In the evening they have soup or tea, white bread, butter, and extras. During the day they are allowed half a bottle of beer with black bread. Wine is given only by medical direction.

The second class are allowed one attendant and a room with varnished furniture to every two patients. Their daily meals include two breakfasts, as enumerated for the first class, with the exception of meat and cheese. At noon and in the afternoon they again partake of food similar to that for the first class. In the evening they have soup or tea with bread and butter. They also receive daily allowances of beer and bread. For these classes there is, on the men's side, a room with verandah opening into an adjoining garden. Here are also provided smoking-room, piano-room, billiard-room, and bath-room. On the female side there is a sitting-room with a verandah opening into a garden. The other accommodations here include a ladies' saloon with piano, a small library, and also facilities for bathing. There is, besides, a dining-room capable of accommodating thirty persons, for the common use of ladies and gentlemen.

The third class have separate accommodation in the principal building, and sleep in associated dormitories containing from three to six beds each. There is one attendant to every ten or twelve patients. Their first breakfast consists of an allowance of coffee with cream and sugar, and white bread; their second breakfast, of warm beer or tea, with bread and butter. At noon they have soup, with rice or groats; four times a week, boiled, and three times roasted meat varied with fish. Potatoes are provided every day, and three times a week other vegetables. In the afternoon they receive coffee with cream and sugar, and the usual allowance of bread and butter. In the evening, tea or soup is served with black bread.

The fourth class occupy dormitories containing from three to seventeen beds each. Their dining-rooms are on the associate plan. There is one attendant to every fifteen patients. Their first breakfast consists of coffee with milk and brown sugar, and an allowance of white bread. The second breakfast for the men includes beer, bread and butter; for the women, warm beer, bread and butter. At noon this class partake of soup similar to that for the third class. Six times a week they have meat; once a week, herrings; four times a week, vegetables with pulse or coddled grains; and potatoes every day. In the afternoon they have coffee as at the first breakfast, with bread and butter. In the evening they sit down to tea or soup, with black bread.

This institution is specially designed for the care of patients having their residence in Hamburg and suburban districts. So far as accommodation permits, however, patients of the first three classes are received from other localities. The ordinary charge covers board, lodging, care, medical attendance, supervision, etc. An advance of four weeks' maintenance is required from the first three classes, besides a deposit for extras, such as excursions, cigars, tobacco, repairs, and washing. The patients are obliged to bring with them clothing and linen plainly marked. A duplicate list of these articles is made out, one of which is returned at the time of committal by way of receipt. When suitable and necessary clothing is not furnished, the officers of the institution are authorized to obtain the same and charge it to those responsible for maintenance. If valuables are intrusted to patients by wish of friends and sanction of the physician, the institution gives no guarantee of their safety. Letters to and from inmates pass through the hands of the chief physician.

The day-rooms in the principal building are heated by Dutch tile-stoves. The corridors are warmed by hot-water pipes. The lighting is by gas. From each of the day-rooms is a passage leading to a small garden or court. For purposes of amusement there are provided a bowling-green, billiard-room, and a large music-hall with pianos. In all the departments of the female side there were pianos. On Sundays, the patients, with consent of the chief physician, attend divine service in the chapel of the asylum, which is provided with separate entrances and accommodation for the male and female patients.

The total number of attendants was given as 101, or 51 men and 50 women. The men enter at a salary of eighteen marks[1] per month, rising at the rate of three marks every two months until the sum of thirty marks is reached. The

[1] A mark (100 pfennings) is equivalent to about twenty-four cents.

women attendants enter at fifteen marks, rising to twenty-four marks per month. There are in addition nine principal *gardes*, each of whom receives forty-two marks per month, and twenty-six female *gardes* at thirty-six marks per month.

In the first, second, and third departments ordinary window sashes hung on hinges are in use. The panes measure 10 x 14 inches. The solitary rooms are variously planned. The blankets used in them were of strong material and double hemmed. The windows of some of these rooms are high in the wall; some are protected by wire screen-work; others again are grated with heavy iron bars. The latter were noticeable in one of the sub-departments for the most violent. In the refractory wards the patients seemed greatly disturbed. Here the glass was of unusual thickness. There are asphalt floors in the rooms occupied by the very filthy.

A considerable number of the chronic insane are provided for in eight two-story houses or pavilions, plainly and cheaply constructed of brick. In connection with each is a small yard. Each of these buildings contains sixty beds. Four *gardes* sleep in each house on the same floor with the patients. The window panes are of a goodly size. Food is brought here from a general kitchen, where it is cooked by steam. The buildings are in the midst of beautiful grounds; but the spectacle of patients restricted to small yards produces an unfavorable impression on the mind of the visitor. As one walks about the grounds and sees the patients through iron screen-work or wooden palings, the idea of a park menagerie is disagreeably suggested. It was asserted by the management that there was no "forcing," and that no mechanical restraint whatever was resorted to.

About one hundred inmates were at work on the farm and in the garden. The farm was stocked with cows, horses, swine, poultry, etc., in the care of which the patients assist. The barns and sheds are conveniently planned and spacious, and have large gardens and grounds about them. There are also sheds in which the insane perform certain kinds of work in unfavorable weather. Two patients were seen at work in the tailor's shop, in which, it was said, there were sometimes as many as six employed. Three were busy in the cabinet shop. The other workers comprised four upholsterers, three carpenters, one tinsmith, five painters, and two blacksmiths.

It has already been indicated that there is a classification of the patients into social grades. But the principle of classification is of much wider application than would be gathered from the mere mention of four classes. It extends to the mental and physical condition of the patients. The aim is not only to observe with care the social distinctions and requirements, but also to provide each of the several classes and their numerous divisions with separate accommodations and conveniences—down even to a separate tea-kitchen. But so complex is this method of classification, that it is admittedly a source of much embarrassment to the administration, and, in consequence, it is carried out with difficulty. The higher grades do not care to be brought in contact with those whom they consider beneath them, while the humbler classes become envious of the better treatment of those ranking higher in the social scale, and the poorer patients, seeing others apparently favored, make demands that cannot be gratified.

The construction of the buildings is, as far as possible, adapted to this principle of classification. On both the men's side and the women's side are divisions for the different classes—each in a sort of separate dwelling with distinct living and sleeping accommodations. The gardens are generally well kept, and the separate tea-kitchens referred to, in almost every instance, open out upon small parterres. In all the principal parts of the institution there was an abundance of blossoming plants, and the pleasing combination of outdoor and indoor attractions appeared to have a beneficial influence on the inmates.

PROVINCIAL INSANE ASYLUM—HALLE, PRUSSIA.

This asylum, which is the older of the two institutions for the insane of the Province of Saxony, has accommodation for 600 patients. On the day of my visit the inmates numbered 585, of whom 304 were men. Like that of Alt-Scherbitz, the asylum receives both acute and chronic cases. Its government is also similar. It is located in the country near Halle, and being an old institution, the buildings, apart from the new hospital, suggest little that would be of advantage to those in sympathy with modern progress.

The land attached to the asylum does not exceed sixty-nine English acres, but it is brought under high cultivation by the inmates. On the day of my visit 111 of the men were engaged in farming, gardening, and other outdoor work; 18 were working as tailors, carpenters, masons, and blacksmiths; and the same number were doing housework. Of the women, 40 were engaged in sewing; 80 at washing, peeling potatoes, and other domestic work; and 12 were employed in the garden. Extensive changes and repairs had been undertaken. Work was progressing for the introduction of a new system to utilize the sewage. It was intended to distribute it by means of narrow trenches over a tract of about four acres set apart for garden purposes. All gas-pipes and water-pipes were to be inclosed within a large conduit, so as to render them more accessible for repairs. The red bricks used for building were of unusual size, being 10 inches long, 4⅝ inches wide, and 2¾ inches thick. They cost thirty-two marks, or nearly eight dollars per thousand. In a cottage connected with the farm resided twenty quiet patients, who worked in the fields under conditions of greater freedom than the others.

An interesting department of the asylum is an infirmary for female patients. This detached, one-story building is capable of accommodating twenty persons, allowing 1,420 cubic feet of air-space to each inmate. The bedsteads had woven-wire bottoms, and were provided with comfortable mattresses. The ventilation, heating, and all sanitary arrangements were quite complete. The windows and doors were without fastenings. The warden's room has a large window overlooking the whole of the interior, so that constant watchfulness may be maintained. The temperature of the body of each patient in the infirmary is taken twice a day and recorded.

In the dormitories of the asylum proper, the beds were provided with adjustable head and foot boards. First-class patients have single rooms; most of the other inmates occupy associated dormitories.

In the dining-room, at the close of the meal one of the patients rose and reverently returned thanks.

BURGHÖLZLI CANTONAL ASYLUM—SWITZERLAND.

This asylum, opened in 1870, is situated amidst beautiful Alpine scenery at Burghölzli, about three miles from Zurich. It is under the immediate direction of a resident Medical Superintendent, who is aided by three assistant physicians. The front or central building is flanked on either side by wings of three stories, all of stone. There is a detached building accommodating a limited number of the chronic insane.

The asylum is designed to accommodate 260 patients, but at the time of my visit it was overcrowded. Patients are admitted to the asylum upon the certificate of a reputable physician, the consent of a near relative or guardian having been obtained. There must also be a guarantee for payment of the charge for support. Recent cases are admitted in preference to chronic, the institution being intended for a hospital for the treatment of acute insanity; while the asylum at Rheinau is especially provided for chronic cases. The two establishments constitute the public provision made by the canton of Zurich for its insane. The inmates are divided into three classes according to payment. Patients sent by the authorities are charged sixty-three centimes per day. Connected with the asylum are seventy acres of land. Every department has its airing-court with high walls. The yards for first-class patients are in front and are laid out in walks and embellished with flowers. The windows of the buildings have outside ornamental gratings and sashes opening on hinges.

In the third-class dining-halls were solid oak tables, benches without backs, and dishes of glazed iron. Tablecloths were in use. First-class patients usually take their meals in their own rooms.

The dormitories were furnished with large-sized wooden French bedsteads, having slats with coiled steel springs at either end. For epileptics were provided special bedsteads having cushioned sideboards to prevent the patients from falling out or otherwise injuring themselves. For those showing a disposition to tear their clothing the beds are covered with oil sail-cloth. First-class patients have single rooms; second-class, quite small associated dormitories; while the third-class occupy apartments containing from six to ten beds.

Chapel accommodation is provided, and religious services are held every Sunday. There is also an amusement hall.

For violent patients who need special treatment, dresses of sail-cloth secured at the back are used, also muffs in exceptional cases. In each refractory ward there were ten isolating cells. These were of a goodly size, but the ventilation was imperfect. The windows were about eight feet above the floor, and the doors were very heavy and solid. There appeared to be a lack of care of the refractory class. In a cell in the female department, for a patient whose habits were said to be very destructive, there was nothing but a loose heap of straw for a bed. Many of the patients seemed violent and excited. It was stated that during the twenty-four hours preceding my visit, one male patient had been isolated for three hours, and one female for five hours. For the same period thirty women were recorded as having been violent in the daytime and eighteen during the night, while there were six men violent by day and seven by night. Fifteen men and eight women were allowed the privilege of walking in the park.

A peculiarity in the treatment here is the mode of giving hot-water baths to refractory patients. The bath-tubs have wooden lids, in which are circular apertures for the neck. The body and limbs are concealed in the tub, from which only the head protrudes, the lid being secured by iron locks. The water is heated to the temperature of 95° F. Under what is termed the prolonged or permanent bath, patients are kept in water from two to ten hours continuously, this treatment being intended to act as a sedative. In such cases the patients are fed in the bath. The writer saw two men and one woman undergoing this treatment. One of the former indicated great exhaustion by a feeble movement of his head; the other two were violent and vociferous. The scene left a painful impression. The day's report showed one man to have been immersed in this way for three hours, another seven hours, another ten hours, and one woman eight hours.

The closets are arranged for the use of buckets, the contents of which are utilized on the farm.

There are two attendants to every fifteen inmates. A male attendant acts as night-watch in the men's department, a female attendant performs a like duty in the women's section, and there is a general night-watch for the whole building. The house is heated by steam.

The industrial system is indicated by the following statement: Seventy-eight women were employed at general domestic work; sixteen were found knitting and sewing in a workroom. Twenty-two men were at work on the farm, three at trades, and fifteen at domestic work. The laundry washing is done mostly by hand, thus giving employment to some of the excitable patients.

The increase of insanity in Switzerland was in a large measure attributed to "hereditary tendency, intemperance, inter-marriage, bad nourishment and consequent anæmia."

LA SALPÊTRIÈRE—PARIS.

This is one of the vast alms-house receptacles for the paupers of Paris, and it is set apart exclusively for females. The buildings, with surrounding gardens, courts, and promenades, occupy a space of seventy-four acres. The whole is inclosed by a high brick wall. The enormous size of the institution may be conceived from the fact that the buildings altogether comprise forty-five large blocks, and are lighted by more than five thousand five hundred windows. Here are the aged infirm, as well as defective children, insane persons, epileptics, and idiots. For the children there is a special section, and in connection therewith a school. At the time of my visit the total population amounted to 6,311, of whom 590 were of unsound mind. Two sections are occupied by the insane ; and these, together with a section for idiots, constitute one of the departments of this institution. There is one attendant to eight or ten of the insane.

In a large yard with gravelled walks and shady trees, there are twelve small brick buildings, with broad projecting roofs like those of Swiss chalets. Each interior is a single apartment fourteen feet square. These cottages, furnished with sliding doors, were used for a peculiar class of the disturbed insane. The clean-washed floors were of oak. The beds had wire springs and abundant bedding. In each cottage were two moderate-sized strongly grated windows with hinged sashes. Heating was effected from below, two furnaces being sufficient to warm the cottages.

For filthy cases, the large associated dormitories had iron bedsteads of the box pattern, with drawers underneath. Thick straw mattresses were in use, and in some cases oil-cloth formed part of the bed furnishing.

There were numerous cells in the insane department. These had two openings, one into an airing-court and another into a corridor. Twelve feeble imbecile women, lightly secured in arm-chairs, were seen in one of the day-rooms. In the bathing department there was a douche, and the tubs were arranged in separate stalls.

Provision is made for entertainments, which include music and dancing. During my visit, a large number of patients were seen in one of the apartments listening to music from a piano-forte.

In one immense kitchen the cooking is done for the whole establishment. The insane eat in a special dining-room. It was said that wine is allowed at each meal, also grapes in their season. Both of these were seen on the tables. An attractive new dining-hall was in course of erection.

This being a very old institution, it has little to engage the attention of the inquirer after modern methods, though it is interesting as showing the kind of provision that a great metropolis has made for the care of some of its dependent classes. Besides, one is curious to see a place noted for its associations with the great reformer Pinel. This is a companion institution to the historically famous Bicêtre. The latter is for male patients, and is similarly managed.

THE ASYLUM OF STE. ANNE—PARIS.

By many French specialists in the treatment of mental disease, this is considered a model institution. It is situated near the southern boundary of Paris, is of modern construction, and, like many other public edifices of the city, is of cut stone. It is built in sections with connecting corridors and intervening courts, and is plain in its exterior. At the time of my visit, the patients numbered upwards of 900, and the sexes were about equally divided.

All classes of the insane are received at Ste. Anne, though the institution is in no sense an asylum, but a reception house and hospital for the treatment of acute insanity. If, from a careful diagnosis, the patient, upon admission, is found to be suffering from acute insanity, he is retained for treatment; if his disease has become chronic, he is transferred to some other asylum, such as Vaucluse, Ville Vraz, or Burge. The hospital is supported entirely by the Department of the Seine, and no paying patients are received, the aim being to relieve those who are unable to pay for skilful treatment. One is impressed with the feeling that the medical idea here predominates. The staff is large, and there is thorough clinical instruction by physicians from the Paris Faculty of Medicine. The Sisters of St. Joseph, of whom there were forty-two, assisted by day servants, had charge of the female side. Each received for her services twenty-five francs per month. In the wards for quiet patients there were three attendants to fifty patients ; in the ward for filthy cases, four attendants to fifty patients ; and in the refractory ward, four attendants to every twelve patients.

The sleeping accommodation is on the second floor. On the lower floor are dining, sitting, and other rooms for day use. The rooms of the refractory wards open upon irregular courts, the walls of which are partially sunk so as to afford glimpses of the country. The isolating cells are made very secure ; some have window-blinds. The bedsteads in these cells were of iron strongly secured to the floor, and were made with a central depression and outlet. Some restraining chairs for the violent were observed placed against the wall. There was one padded room for men and another for women. The airing-courts, which contained many noisy patients, were cheerless and quite limited in size. In connection with the infirmary is a small court, in which were flowers and shrubbery. Roofed galleries are provided, permitting exercise without exposure. The various wards are heated by means of pipes laid under perforated iron plates in the floor.

In the bath-room the tubs were not only separated from each other, but set away from the walls. There were here a plunge bath, a hose for use upon the patients, and a douche with arrangement for securing the person while undergoing treatment. In one of the rooms of the bathing department were stationary glazed iron foot-baths. There were, besides, arrangements for vapor baths. A room was specially set apart for washing and shampooing after the use of the vapor bath. As in some other French institutions, hydropathy has a prominent place in the treatment, and there is much reliance placed upon it as a means of allaying excitement.

The wards of this institution are small, which is desirable in the treatment of acute cases. Many of the apartments, however, are poorly lighted from small airing-courts, and all are quite plainly furnished. There appeared to be a lack of air, light, and room, and the provision generally did not seem well adapted to secure the best results in curative treatment.

INSANE ASYLUM AT CHARENTON.

This, one of the very old institutions of France, founded in 1642, is in the suburbs of Paris, near the park of Vincennes, and receives, as do French institutions generally, curable and incurable patients. Charenton has had a troublous history, having been several times destroyed and successively restored. It is said that at one time it was a monastery belonging to the monks of St. Jean de Dieu, who here undertook the care of the insane. Following the revolution of 1830 it became, under Louis Phillippe, the property of the State, and was at that time rebuilt of white limestone pretty much as it now stands. The place has interesting associations from its connection with the names of Esquirol and other distinguished alienists who here practised their profession and wrote many of their valuable works.

The buildings occupy a commanding site on a finely terraced slope. From the chapel, which is above the other structures, is obtained a view of a charming landscape stretching along the course of the Seine. This edifice accommodates about one hundred persons. The terraced airing-courts, of which there are eight in the male and an equal number in the female department, measure each from one hundred to two hundred feet square, and have covered places for shelter on either side. From these courts, made cool by shade and beautified by fountains and flowers, there are delightful views, which must have a salutary effect on the inmates. Besides these courts, there are spacious grounds shaded by large trees. In these open spaces, at certain hours of the day in favorable weather, all save the refractory take recreation.

On the day of my visit, the asylum contained 285 men and 300 women. Most of the patients are from the middle class of society and are supported principally by relatives or friends. There are two resident physicians, two medical assistants, a consulting surgeon, and an apothecary. A resident chaplain of the Roman Catholic Church daily conducts religious services for such patients as desire to attend. Subordinate to the physician, the female department is under the charge of the Sisters *religieuse* of the order of St. Augustine, who act as attendants and are assisted by servants. The female servants wear blue dresses with white caps and aprons. Each of the Sisters receives twenty-five francs per month.

That portion of the institution occupied by the patients is but two stories high. The windows of the upper floor have gratings with five-inch square spaces: those of the ground-floor are without guards. The asylum is planned so as to admit plenty of light. The buildings are warmed by hot-water pipes.

Each of the dining-rooms has ten tables, each table accommodating six persons. Cushioned benches without backs were used instead of chairs. The table furnishing included crockery, wine-glasses, water-bottles, spoons, forks, etc. In the refractory ward, no knives were used, and its dining-room benches were uncushioned. Wine is supplied at all the meals. The diet is regulated by the price paid.

The dormitories are on the second floor. Iron bedsteads measuring about four feet in width were in use. The mattresses, made of horse-hair and wool, were laid on coil springs, and the bedding was good and plentiful. Patients requiring special provision had deep box-bedsteads filled with marine grass. These had outlets in the centre. The uncleanly occupied a separate one-story building, and had also a separate airing-court. In the dormitories for women, over the beds of the quieter patients were pretty dimity canopies.

In the bath-rooms, the tubs are arranged so as to be accessible from all sides. As at Ste. Anne's, bathing is an important factor in treatment. The baths include medicated, Turkish, and Roman. There are also douches and packs.

A comfortably furnished room, in which were two long tables, was provided with cushioned benches and chairs at each side. This apartment is used by the male patients as a reading-room. The library is valuable, and includes the books and writings of Esquirol, which he presented to the asylum. Besides the collection of medical works, there is an abundance of entertaining reading in the form of books, papers, and magazines. There are no less than three billiard-rooms. In the corridors adjoining these places of recreation, a light was kept burning for the accommodation of smokers. There is also a spacious amusement-room in which the patients assemble every Sunday from 7 P.M. to 10 P.M.

The refractory wards communicate with a wide corridor having windows about eight feet above the floor. For the extremely filthy of both sexes, there are small courts for seclusion, and into these open a few isolating cells. Each of the latter has a cemented floor, and is warmed by steam. Box-bedsteads are here used, though in one of the cells on the men's side, in which the atmosphere was very offensive, nothing but straw was seen for a bed, and the floor and walls were besmeared with filth, presenting an unsightly appearance. Strait-jackets, it was stated, were used on an average four or five times a day. The restraining chairs, three of which were occupied at the time of my visit, have cushioned straps for the feet. There were no cribs of any kind. Even the movable ones, formerly used for carrying patients about, had been discarded. It was asserted here, as elsewhere, that none were likely to be found in France.

No industrial system exists, although inmates who desire to work in the garden or among the flowers are permitted to do so, and light needlework is provided for women.

At the time of my visit patients were received into the institution at from 1,000 to 1,400 and even 1,800 francs per annum. The number of attendants average one to about every nine patients. Such of the men as paid 900 francs extra, and those of the women who paid 850 francs extra, were allowed a servant as well as a private room, and ante-room for servant. By the payment of a still higher sum, a patient was entitled to a suite of apartments in addition to a servant's room. The furniture, etc., where no extra fees were paid, was nearly uniform. There are about forty single rooms for each sex. These are fitted up somewhat elaborately, with mahogany bedsteads and other corresponding furniture. Many of the apartments have parquet floors and are highly polished with wax.

What seemed remarkable in an institution possessed of so many comforts and having accommodation for so many private patients, was the fact that it had few pictures on its walls, though a tasteful display of bric-a-brac was noticed in some of the rooms for women.

CLERMONT EN OISE—FRANCE.

The asylum at Clermont, organized as a private and industrial institution, has long attracted attention by reason of its successful plan of colonizing the insane. It is situated in the department of the Oise, thirty-six miles north of Paris. Up to the time of my visit it had been managed and superintended by that distinguished alienist, Dr. Gustave Labitte, who was then assisted by his son, Dr. George Labitte.

There were about 1,580 patients under the care of the general management. These were distributed over the three departments of the asylum as follows : One thousand, including twenty epileptic, idiotic, and insane children, at the Clermont central asylum ; about 440 in what is known as the colony of Fitz-James ; and 140 in a second colony at Villers. The central office at Clermont is the headquarters of the superintendent, and is in telephonic communication with Fitz-James and Villers. Besides receiving insane poor from several outlying departments or districts, this institution admits boarders on terms arranged by special contract. Of the total number of patients, it was estimated that about 319 were boarders and the remainder paupers.

For the maintenance of the pauper insane the authorities paid, at the time of my visit, at the rate of one franc, twenty centimes per day. The indigent insane supported by their friends were received at the same rate. Boarders having private rooms and better fare paid 250 francs per month, or, if they had their own servant, 350 francs per month. Single bedrooms were allowed to those who paid the higher prices, and associated dormitories with from five to sixteen beds were provided for those who paid the lower rates. The boarders, many of whom live at Fitz-James, do no work.

Taking the three departments in order, we begin with the central asylum at Clermont, connected with which are some fifty acres of improved and garden ground. Both sexes are here cared for, a strict separation being maintained. Among the inmates of the male department were about fifty epileptics, between eighty and ninety sick and filthy, and 150 quiet patients, from the last of whom laborers for the grounds are selected. Each class occupies a separate section provided with associate apartments for sleeping and eating. The arrangements for female inmates are substantially the same as those for men. For the quiet class in the male department there are three attendants to one hundred and fifty patients; for the refractory, four to forty; and for the twenty children, two attendants. The sleeping apartments are so arranged that the attendants can render prompt assistance in case of emergency. Patients paid for by the authorities have different sleeping accommodation from that of indigent patients supported by their friends. The former sleep in associated dormitories containing forty-five beds each; the latter occupy smaller apartments, each having sixteen beds.

In the wards for filthy patients, the bedsteads were of the deep box pattern. The sheets were laid directly on sea grass, which was changed daily. In this way, it was said, the beds were more easily and cheaply kept clean. The metallic bottom had a depression in the centre with outlet. A breadth of carpet was laid on the tile floor beside each bed. The questionable practice of keeping uncleanly and violent patients much of the time in bed prevailed here.

The infirmary on the second floor, in which are kept all suspected of suicidal tendencies, as well as the violent requiring special restraint, has grated windows. The iron bars are five and a half inches apart. Ordinarily, the windows have four lockable sashes, the two smaller ones being at the top; and all are hung on hinges. In this portion of the building the bedsteads were of iron. For restraining excited patients a peculiar method, and one not confined to Clermont, is practised. The patient is attired in a strong dress with long sleeves and legs, the extremities of which are secured by cloth bands to the side and foot boards of the bedstead. A band similar to these is attached to the collar of the dress and fastened to the head-board, thus holding the patient in a recumbent position.

There are ten cells for isolation in the whole establishment, all of which, at the time of my visit, were unoccupied. They are much like the other rooms, except that they are made secure by wooden blinds. The floors are of oak and waxed. The bedsteads here, as in the infirmary, were of iron, box pattern, with holes at the head, foot, and sides for fastening restraining bands when they are considered necessary. There were no padded cells. There were two restraining chairs, which were said to be seldom used. Narcotics were sometimes given to allay excitement. The strait-jacket or camisole was used in cases of extreme violence.

The dormitory of a refractory ward that I inspected had grated windows. The pine floors were neatly waxed; the bedsteads were of iron. Two attendants slept in an apartment adjoining and overlooking the ward.

On one side of a room on the first floor, the windows are placed high in the wall. On the opposite side, opening to the adjoining court, the windows are lower and double the size of the others. The furniture was of the plainest description; the tables and benches were deal. Adjoining is a small dormitory for such as are unable to go up-stairs. The objectionable arrangement of high windows already described was seen in a building in course of erection.

The airing-courts are surrounded by high walls and are thus entirely separated one from the other. They are shaded by trees, and generally measure from 200 to 300 feet square. The airing-court for boarders on the female side is very large, well laid out, and ornamented with flowers and shrubbery. It is divided by railings into three portions, occupied

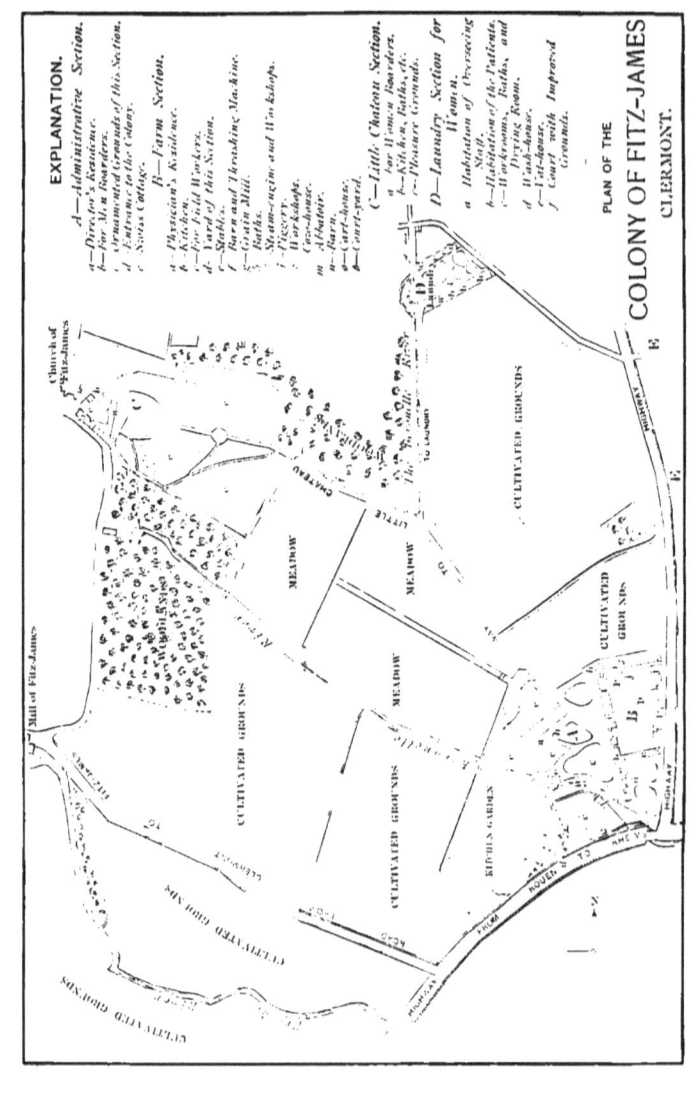

by the quiet, the unquiet or refractory, and the uncleanly. In the centre of that part of the court which is nearest to the main building, is a detached open pavilion used by the women patients for sewing and reading. In the division for the filthy class, some of the patients were standing or walking about, some sitting, some squatting, and some even lying on the pavement. There was here a general appearance of untidiness and disorder, doubtless attributable to a lack of attendants.

Some new two-story buildings were in course of erection; the upper portions being intended for dormitory purposes and the ground-floor solely for day use.

The institution dress for men in summer includes a blue blouse worn over a jacket of the same color, dark woollen pantaloons, and shoes with wooden soles. In winter a warm jacket or coat is added. To all except boarders clothing is furnished by the management.

At the central asylum of Clermont, it was said that the inmates were employed as far as practicable. The women sew for the institution. The men are chiefly occupied in vegetable gardening. Two were employed as coopers, four as cabinet-makers and joiners, four as painters, two as tinsmiths, seven as bakers, twelve as shoemakers, six as mattress-makers, and a few were engaged in making general repairs. Reading and assembly rooms are connected with the institution.

The colony of Fitz-James, which is about a mile distant from Clermont, presents to the stranger many points of interest. Industry, order, and freedom are noticeable throughout the whole establishment. There are no enclosing walls. The two-story buildings of the administrative section, surrounded by gardens, parks, and meadows, are plain, and are planned so as to afford sleeping apartments above and day-rooms below. Prominent among these buildings is one allotted to boarders, in which are a billiard-room and a library. Here was also an organ. All the apartments in this building, including the dining and day rooms, were comfortably and tastefully, though not luxuriously furnished. The day and sitting rooms had plush-covered ottomans, cushioned chairs, and other upholstered furniture. The number of attendants was large. The apartments are arranged so as to admit of servants sleeping in rooms communicating on either side with the boarders on whom they wait.

The insane poor sleep in associated dormitories, containing from twenty to thirty beds each. There are no interior partitions, and the windows, which are opposite each other, are kept open not only in summer, but also in winter. The bedsteads were of iron. The bedding consisted of one straw and one woollen mattress, linen sheets, and woollen blankets. In every case the beds were neatly made up. The clean waxed floors added to the tidy appearance of the apartments.

A short walk from the main buildings brings one to an establishment called the " Little Château." Here, under conditions of much freedom, a number of " lady boarders " are provided for, in accordance with the requirements of their individual preferences and temperaments.

In another direction from the administrative section, and forming a somewhat remarkable colony by itself, is the large laundry (*la section de Bécrel*), where the patients engaged at laundrying reside. The laundry buildings are arranged on two sides of a court, and the women are effectually secluded. This interior space is large, and made pleasant with flowers and greensward. The rooms are well lighted. They were comfortably though plainly furnished, and tidily kept. There was a variety of wholesome food. The dining-room tables were of deal. Earthen dishes, glasses, and tin spoons were used. The laundry machinery is propelled by steam. Here all the washing of the establishment is done. A novel arrangement in laundrying is found in this department. In a very large room with stone floor is a long and broad cement vat, through which flows a stream of pure soft water that has been turned from its course for service here. Those engaged in washing are ranged on either side of this immense vat, which is at a convenient elevation from the floor. The different articles are first laid on the marginal slope of the trough, rubbed with soap, and then washed and rinsed in the flowing stream. The clothes, when practicable, are dried in the open air. The most violent of the female patients who cannot be kept at work that does not require muscular exertion are engaged in washing by hand; others less excitable are employed at the machines, or in spreading the linen to dry; while the more delicate are occupied with folding and smoothing the linen. The grounds here are sufficiently extensive to afford recreation and considerable freedom; at the same time there is the seclusion necessary for the protection of the insane women.

The colonies of Fitz-James and Villers have upwards of a thousand acres of farming land. The majority of the male pauper patients are employed at general farm-work. They are divided into groups of ten or fifteen, and are directed and assisted by an attendant. Among so large a number there may be found those adapted to every requirement of farm industry. By carefully studying the tastes, disposition, and ability of each patient, selection is made of those to whom can be assigned a specific charge, such as the care of certain tools. In this way the land is thoroughly cultivated, the farm implements kept in repair, and the domestic animals properly cared for. Vegetables in great variety are produced in quantities sufficient to meet the wants of the institution. Dr. Labitte says: "We have been able for many years to raise upon the institution grounds, with the exception of an occasional purchase, all our cattle and horses." In the capacious farm buildings and stock-yards, which take up five acres, were thirty horses and a large herd of cows and growing stock. In the piggery were upwards of a hundred swine. Not a little attention is given here to the breeding of rabbits. Great pains is taken to secure the best breeds of farm stock, and the care bestowed upon it is often rewarded by prizes taken at exhibitions and fairs. Dr. Labitte says: "The insane patient who lives in the midst of this progress is proud of the prizes that the farm receives each year, and of which he can attribute to himself a share. He shows complacently to visitors the animals he has raised and for which he cares." In connection with the farm buildings at Fitz-James there is accommodation for a small colony of farm laborers, for whom have been fitted up day-rooms on the ground-floor, and associated dormitories above. As one enters the court around which are arranged the stables, sheep-folds, cow-sheds, and piggeries, he is struck with the neatness and order with which they are kept. Among the attendants are several practical farmers who work with and direct the insane. About the barns was seen a large group of patients cheerfully preparing for work after their mid-day meal. One in a gleeful mood was setting out for the field mounted on one of the work-horses.

The sewage, after being conveyed to vats in the fields, passes successively through four of them, which are placed at different elevations for the purpose of separating the liquid from the solid. This Dr. Labitte regarded as the best-known system of sewage utilization in connection with asylums.

But agriculture is not the only industry of Fitz-James.

While many of the patients are occupied on the land, there are a considerable number chosen, on account of previous experience, to work at various trades. On the farm are two mills for grinding wheat, a forge and a machine-shop for repairing agricultural implements, a mill used in making cement, a cider manufactory, a bakery, an abattoir, and a dairy. Each of the flouring mills has three run of stone. One is propelled by an engine of thirty horse-power, and the other by water. These mills, especially the one having the steam-engine, were shown with much pride. In the blacksmith's shop were long rows of horseshoes, all made by the patients.

In order to make the classification still more perfect, simplify attendance, and give the greatest amount of freedom to each class, there are selected for the colony of Villers only chronic male patients who require little oversight and are already accustomed to farm-work. Although the colony of Villers is a considerable distance from that of Fitz-James, the lands adjoin. The open-door system is here fully realized, the operations of the place being conducted with nearly the same liberty as in ordinary family life.

On Sunday many of the patients attend mass in the village, where a portion of the church is set apart for them. Some are trusted to go alone both to church and to village festivals.

The examination of the Clermont colonies afforded me much satisfaction. There was apparent cheerfulness in the several divisions, while, as has already been shown, the steadiness with which the industries are pursued tells favorably in the economic results attained. Nevertheless, the primary object of labor is the welfare of the patients. The force of example, the fear of being returned to the more restricted limits of the central asylum, and the offer of rewards in the shape of money or dainty food are inducements to labor; but recourse to coercion is forbidden. On an average, the time allotted for work does not extend beyond six hours per day. In some departments, as, for example, the laundry, the labor is performed by alternate sets of workers. The free and natural conditions of life existing at Fitz-James and Villers are marked characteristics of these colonies, nor can one forbear to note the admirable judgment and delicate tact displayed in adjusting the employments to the experience, physical capacity, and mental condition of the patient. If a working patient betrays symptoms of violence or becomes from any cause unmanageable, he is immediately transferred to the central asylum; and, on the other hand, whenever a patient in the central asylum becomes quiet and tractable, he is removed to the colony and its industrial system. There is thus a steady and highly beneficial intercourse between the two divisions. With all the freedom here accorded, surprising as it may appear, there are few attempts to escape, and, indeed, there are few casualties of any description.

The satisfaction derived from the inspection of the colonies of Fitz-James and Villers did not extend to the central institution with its one thousand inmates. Possibly owing to the large numbers brought under one management and the burdens incident to the conducting of its extended business affairs, the central asylum at Clermont did not, as it appeared to the writer, reach a proper standard for a hospital for the treatment of the acute insane.

CHAPTER VI.

THE COLONY OF GHEEL.

LEAVING the afternoon train from Antwerp, the writer, accompanied by a stenographer familiar with the patois of the place, wended his way for about half a mile along a street or roadway sparsely lined with plain cottages to one of the two principal inns of Gheel. A few steps from this humble hostlery brought us to the main public square, at one end of which is the church of St. Amand, and close by is a lofty cross with a life-size figure of our Saviour. On the remaining three sides of the square are plain houses or shops, none of which exceed two stories in height. Near the church is a huge ancient-looking pump which forms a rendezvous for the young maids and children of the town, as well as a centre for the dissemination of gossipy news. The farther end of the green was covered with linen of the villagers spread out with great particularity to bleach upon the grass, indicating the methodical habits of the people. A number of persons having the appearance of being insane were seen walking along the street or seated on the door-steps of the dwellings. From this part of the staid old town, we proceeded along a winding paved road, on either side of which were brick houses with red tile roofs, receiving on our way respectful salutations from the villagers. A novel street scene was that of men in small cars which were drawn by dogs. The first car noticed had five of these animals abreast; to each of two others was attached a pair of dogs. Thus far we had passed a number of small shops devoted to the sale of beer and tobacco. About an eighth of a mile from the square, the street terminates in a long avenue of beautiful elms, extending far into the country, with here and there, on either side, larger two-story buildings than we had previously seen. A short walk along this avenue brought us to the Infirmary, a large two-story brick structure with red tile roof, standing amid gardens and shrubbery. Occupying a niche in the central projecting gable of this building is a life-size figure of St. Dymphna, the patron saint of Gheel. Having made an appointment for the following morning with Dr. Peeters, the head physician and director of this institution, we took a short route across the cultivated patches of ground and soon reached the church of St. Dymphna, of legendary fame.

This church is a massive cruciform structure, chiefly of brick, with portals and carved mouldings of stone. Within are numerous tombs and tablets, and the walls have paintings and other decorations in keeping with the sacred character of the place. As we entered, through the gathering shades of evening, we descried at the farther extremity a star-like light which burns unceasingly at the shrine of the saint. On one of the church pillars, near the altar, is a canopied statue of St. Dymphna, and opposite to it on the corresponding pillar, is a sculptured figure of Father Gerebernus, the religious guide and protector of the maiden saint. The altar of white marble has sculptured representations of insane persons with manacled limbs, in the act of supplicating St. Dymphna. The silver shrine, believed to contain her relics, is elaborately wrought, and is preserved with religious care. The tomb of the saint is elevated on short pillars, but sufficiently high to admit of persons passing under it upon their knees. The stone pavement about it is much worn by the feet of weary pilgrims, who, through many

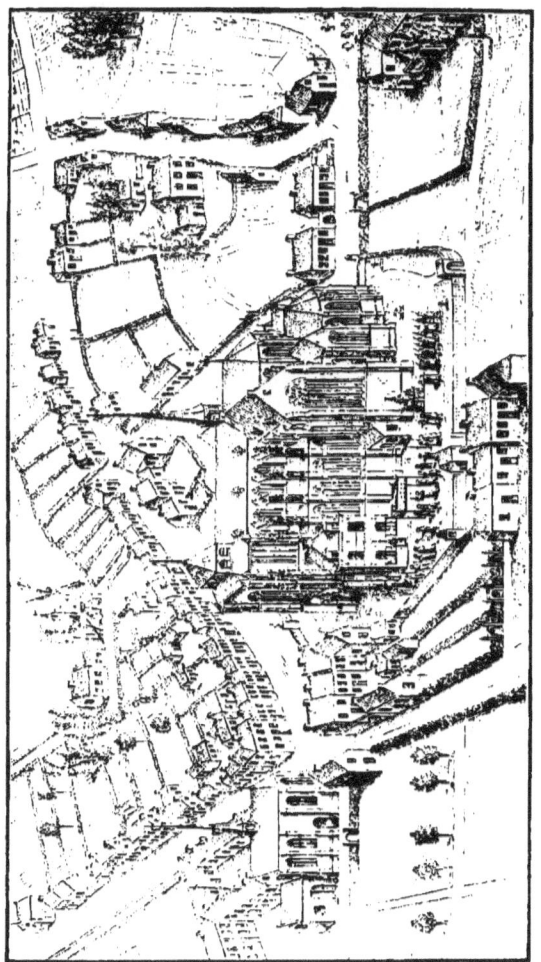

THE CHURCH OF ST. DYMPHNA, GHEEL.
From an old picture

generations, have here sought relief from their afflictions. On the walls near by are legendary inscriptions relating to her romantic life.

In apartments connected with the church and under the charge of a resident sacristan is a goodly sized room with tile floor, and massive beams overhead. On each side of a large open fire-place are iron rings in the wall once used to secure violent patients near the fire without endangering them. Adjoining this room is one not so large, which receives light from the other through a small grated window. The smaller room, unoccupied at the time of my visit, and now seldom used, was furnished with a table and a comfortable bed. According to Esquirol, who visited Gheel in the year 1821 : " It was formerly the custom for a patient to be placed, on his arrival, for nine successive days in the house adjoining the church, along with a few others, in a room under the guard of two old women. A priest came every day and said mass and read prayers to them. Then those patients who were tranquil, joined by some children of the place, went three times in procession outside and as often inside of the church during the nine days of treatment. Each time they arrived at the tomb of the saint they prostrated themselves and passed under it on their knees. If a patient was too furious to allow of this being done, an inhabitant of Gheel and some children were paid to perform the ceremony."

However incongruous these ceremonies may seem as compared with our present theories respecting the insane, we may not treat with levity the simple faith of those, who, actuated by a belief in spiritual influences, sought relief for themselves or suffering friends at the shrine of St. Dymphna. The pious custodian in charge of the church lamented the falling off in the number of pilgrims' visits, a circumstance attributed by him in part to " new-fangled notions of the doctors." Though faith in the healing powers of the saint is weakened, her memory is still honored by an annual fête held in the month of May, in which both sane and insane take part.

Not far from the church of St. Dymphna stands an ancient hospital, said to have been built in the thirteenth century. In one of its walls is a covered recess, open to the street, from which it is separated by iron bars. It contains two figures, life-size ; one representing a grim warrior, the other, a kneeling and suppliant maiden upon whom his uplifted sword is descending. These figures, it was said, mark the spot where the saint was put to death.

There are various versions of the interesting legend of St. Dymphna, and among them the following, which is given substantially as related to us by an accredited *cicerone*, who evinced a zealous interest in all that affected the good reputation of his native town.

In the seventh century there lived in Erin a king whose daughter, like her mother, had acquired fame for her beauty as well as her many virtues. She became a convert to Christianity, and shortly after her conversion, her father, tempted by the Evil One, so persecuted his daughter by his unnatural conduct as to drive her to desperation. She took counsel of an aged priest, the holy Gerebernus, and with his advice and assistance secretly left the country and found a temporary refuge at the place where Gheel now stands. Here, under the protection of Father Gerebernus, she abode, serving God by fasting and prayer. The king, on discovering that his daughter had fled, was beside himself with rage, and sought her everywhere in his own kingdom. Unsuccessful in these efforts, he extended his search to other countries, and at length came to Antwerp, where, finding some pieces of the coin of his own realm, they afforded him a clue, which, with bribery, led to the betrayal of her secret abiding-place. Here he found and slew Father Gerebernus, and then, with his own hand, put his lovely daughter to death. God protected their bodies from the wild beasts, and from the fowls of the air, until they were buried by the people. When these things were told about the country, many came, or were brought, to the graves of the martyrs, and were cured of all manner of diseases, bodily as well as mental, but particularly the latter.

The priests, hearing of the miracles that were effected, desired that the bodies should be exhumed, and again buried with appropriate ceremonies. In this attempt, they were astonished to find that the remains of Dymphna and Gerebernus had each been inclosed in a snow-white stone sarcophagus. This was regarded as a special interposition of the angels. Afterward, many came from afar to see the relics of the holy dead. Among these strangers were some inhabitants of Xanten on the Rhine, who, after remaining several days at Gheel, meanly stole the sacred relics and fled, hotly pursued by the indignant people of Gheel. Becoming pressed to make good their escape, the wicked plunderers relinquished one of the sarcophagi, and, by mistake, made off with the remains of Father Gerebernus, leaving the coffin of the virgin in the hands of their pursuers. Their mistake discovered, the sacrilegious villains returned and made an attempt to recapture the coffin of Dymphna, when lo ! it was found to have become as immovable as the rock on which it lay. A good frow, prompted by a heavenly vision, harnessed her young heifers, and, with them, drew the sacred relics back to Gheel, aided by some and mocked by others. It was believed that the martyred maiden not only forgave the murderous deed of her father, believing him to have been insane, but desired to alleviate a like madness in others. Her presence was proclaimed at the spot, and thither afterward pilgrims came imploring her intercession. Miraculous cures were effected, and many of unsound mind were restored to reason at her shrine.

As time wore on, added our guide, pilgrims increasingly flocked to this place, and out of their gifts a chapel was at first erected, and some time in the twelfth century the present church of St. Dymphna. It became necessary for the friends of patients to find in the neighborhood permanent boarding-places for the afflicted, and thus was established the colony of Gheel.

THE SURROUNDINGS OF GHEEL.

The commune, or what is known as the colony of Gheel, covers a territory of upwards of forty square miles. Its greatest length is about eleven miles, and its greatest breadth a little short of nine miles. The colony contains about three thousand houses, in more than one third of which lunatics are boarded. The public affairs of the commune are presided over by a Burgomaster. A sluggish stream, called the Greater Nethe, flows through the flat country, in the midst of which it is situated. The town proper, which partakes of the character of a straggling village, is centrally located in the commune, over which, at intervals, are scattered hamlets, while the whole area of the country is dotted with farm-houses, having very small cultivated farms or garden fields attached. The soil is generally suitable for gardening, and for the most part fairly productive, though in some portions light and sandy. Every available foot of land is under tillage, not even a path along the wayside being left uncultivated, and often, on other than the principal roads, there is scarcely driving space between the fields. The parish church of St. Amand and that of St. Dymphna form two central points of settlement, between which is the main business street of the town, extending a distance of three eighths of a mile. In addition to the long straight avenue previously referred to, there are several others of modern appearance; but aside from these, the whole country about Gheel is a tangled web of roads and lanes, along which, at intervals, are farm cottages.

In no part of the colony is there evidence of opulence, or, indeed, of any thing more than middle-class comfort. Nearly all the towns-people are farmers in a small way, frugal and industrious. Many of them keep little other stock than a couple of cows, and, besides raising enough to supply their own wants, eke out a rent with what they receive for boarding and attending to the insane. The women are not less industrious than the men. It is no uncommon sight to see females at work in the fields, or driving dogs harnessed to cars. It is estimated that some three hundred or four hundred women are engaged in this place at lacemaking, their day's work of twelve hours yielding them barely a franc. In the town are several tobacco factories and two breweries. What was formerly a paper factory, employing one hundred hands in the manufacture of colored paper for the English market, was pointed out to us. It had succumbed to the pressure of rival interests.

It may be well, for purposes of comparison, and for the better comprehension of what follows, to give here a few statistics relating to the insane in Gheel at different periods since the latter part of the eighteenth century, when the number was about four hundred.

Year.	Patients.	Year.	Patients.
1803	600	1870	1,095
1812	500	1871	1,127
1821	400	1873	1,230
1840	717	1874	1,272
1855	778	1876	1,383
1868	1,035	1880	1,630
1869	1,072	1885	1,653

Prior to 1803, there was no organized governmental control, nor any regular system of inspection of the insane. The commune, comprising eight parishes, each having a church, has nearly 12,000 inhabitants, of whom about 5,000 live in the town. The proportion of insane is about one to seven of the whole population of the colony.

THE INFIRMARY AND ITS FUNCTIONS.

The infirmary fulfils the functions of a central asylum, or "observation" station, through which all cases received into the colony must pass, and to which are returned such as are found unsuited to family life, previous to their being sent elsewhere for asylum care. It serves also as a hospital for such of the insane as require medical treatment, and provides for those who are subject to periodical paroxysms. This, as also the whole of the boarding-out system of the colony, is under the direction of a Board of Commissioners, including the Governor of the Province, the Judge of the Canton, the Burgomaster, the King's Procurator, a physician appointed by the Government, and two members chosen by the Minister of Justice. This Commission has supervision of all matters relating to the general care of the patients, and makes a yearly report to the Government. There is, in addition to this, a permanent committee upon which devolves the enforcement of the regulations respecting lunatics in the colony. The commune and the medical staff of the colony are represented upon this committee, of which the Burgomaster is chairman. The committee meet once a week, to decide as to the eligibility of new arrivals, and as to their distribution in the colony. They likewise receive and pay out money, and adjudicate upon the qualifications of all applicants for the boarding of the insane. For the latter purpose a registered list is kept of all who wish to receive patients. The Secretary of the committee and of the Commission also acts as steward of the infirmary.

The district, or commune, is divided into two medical sections, each of which is in charge of a chief physician assisted by a second medical officer. The time of these sectional physicians is not wholly devoted to the insane, as they have also a private practice. The Medical Director of the colony, Dr. Peeters, is charged with the duty of general visitation and inspection. The sectional physicians or their assistants are, by rule, required to visit every one of their patients once a month, and troublesome cases whenever necessary. Acute cases are visited at least once a week. Before 1879 the doctors of the sections were not required to visit their patients oftener than three or four times a year. In addition to the medical supervising officers there are four *gardes*, or inspectors, appointed by the Minister of Justice, each having an assigned district. These are charged with the duty of visiting the patients boarded out and examining their dietary, clothing, and bedding, and inquiring into whatever relates to their proper care or treatment. They also take charge of the transfer of patients to and from the infirmary. In addition to the four district inspectors two others are stationed at the infirmary, who act under the direction of the Medical Superintendent. Each district inspector is required to see every patient within the district assigned him once a fortnight, and he is subject to call at any time, in case of violent outbreaks, or where restraint is necessary. He reports daily to the head physician; and his visits to the patients, as well as those of the medical men, are duly recorded in register-books, which are kept in the houses where the patients board. In these books are inserted particulars of the name, age, etc., of each patient, as well as a note of the clothing received from the infirmary. Besides the officials already mentioned, there are in attendance at the infirmary five Sisters *religieuse* of the order of the Norbertines. There are also three male attendants.

The infirmary is a plain, but substantial structure, with a capacity to accommodate about one hundred inmates. Attached to the institution are five acres of ground. The insane inmates at the time of my visit numbered twenty-six males and twenty females. The two wings of the infirmary are built on a similar plan, each embracing three departments—the first being for " quiet," the second for " partially disturbed," and the third for " disturbed " patients. The men and women occupy different sides of the building, both being supplied with food from a common kitchen. Each department contains a dining-room, small hospital, and inner court. In each of the courts was a large dove-cot, erected on a high, central post, for the entertainment of the insane.

The isolating cells, of which there were seven for men and seven for women, contained wooden bedsteads of box pattern, with head and foot boards, and each was provided with a chair. The floors are of wood. A hall or corridor extends on either side of the range of cells, which are tolerably well-lighted, and warmed by stoves from the outside. One corridor is used for domestic work, the other for observation of the patients. A window, the gratings of which form geometrical figures, opens into one of the halls. There is a solid-panelled door to each cell. Each of these small rooms has openings to facilitate ventilation. In a corner of the room is a tin or iron night-vessel of peculiar pattern, fitting close against the seat, and accessible for removal only from the outside. In the third division there are in addition cells of similar construction to those described, with, however, stronger gratings. There was, on either side, a padded room; but these, it was said, were seldom used.

There were on the men's side some iron bedsteads thirty-three inches wide, with panels grained to resemble wood. Fastened to the bottom of each of these bedsteads was a plate of galvanized iron, about three feet long, with a depression converging to a central aperture leading to a metallic drawer-like receptacle underneath. The side-pieces of the bedstead, ten inches deep, were made of sheet-iron, with the edges strengthened by thin iron bars. On each side were three loops, used, when necessary, in connection with a restraining apparatus. The bottom mattress of the bed was made in three sections.

In connection with a room for the sick was a spacious

porch inclosed with glazed sash, and having a glass roof to admit sunlight.

Dr. Peeters said it had been found impracticable to dispense with mechanical restraint. The appliances in use were as follows: 1. A broad padded belt, or band, five inches wide, for locking around the upper part of the arm. In this belt is an iron loop through which passes a strap for securing the patient to the side of the bed. This, we were told, could be used without accident, however much the patient might struggle. 2. Two padded leather bands three inches broad, connected in the centre by a link two inches long. These are designed for the ankles of the patient. To each is attached a loop of iron through which a strap is passed and fastened to either side of the bedstead. 3. Two padded soft leather wrist-bands, three inches wide, separating the wrists about eight inches. Dr. Peeters said these were scarcely ever used. 4. A broad leather belt for encircling the body, with two leather pockets in front for the hands, which are secured in their position by wristlet-straps. Extending from this belt across the shoulders, suspender fashion, are straps which retain in position and relieve the weight of the belt, which locks at the back. This, familiarly known as *la ceinture à bracelets mobiles*, was said to be preferred by the patients to the camisole, and was spoken of as "particularly useful for such as incline to tear their clothing." It should be stated that I saw no patients wearing any of these contrivances during my inspection of the infirmary.

THE "HÔTES" AND "NOURRICIERS."

The need of some kind of governmental supervision and control over the insane at Gheel became manifest at an early period. From the past history of the colony it appears that, at the beginning of the present century, the "trade" interest of the peasantry, and of the communal administration in the keeping of the insane, assumed an objectionable form. The local administration exercised a sort of favoritism in allocating the patients boarded out, and the favored caretakers, in return, submitted to a tax on their land or business. Eventually, as numbers increased, and the public became more enlightened on the subject, the communes sending patients to Gheel, resolved also to send delegates, or inspectors, charged with the duty of watching over the welfare of those whom the authorities had consigned to colonial care. Dr. Parigot, who was appointed to represent Brussels in this capacity, states that when inspectors were first appointed, in 1803, the treatment of lunatics at Gheel was perhaps even more cruel than that of the negroes he had seen in South America.

Under the legislative reform of 1850 the control and administration of the colony, so far as it related to the insane, passed from the communal authorities to the State government, and the changes then made and those effected by the law of 1851, led eventually to the erection of the infirmary building, which was opened in 1862. Previous to 1851 the system in operation in the colony was not free from reprehensible practices. Severe forms of restraint were then in use. Dr. Pliny Earle relates that, during his visit to Belgium in 1849, he observed "a patient in the streets of Gheel, with his waist encircled in an iron belt, to which his hands were secured by wristlets," and that "in the suburbs, and among the farmhouses, several were fettered with iron, the chain between the ankles being about eight inches in length, while, in some cases, the rings round the ankles had abraded the skin and occasioned bad ulcers." The appointment of Dr. Bulckens as director, in 1856, brought about the gradual abolition of these unnatural methods of restraint. Since then the treatment has been of a more humane character. During the whole period of my visit I did not see a single patient under any form of mechanical restraint.

The colony is open to certain classes of the insane for whose board compensation is made. In the case of private patients the price is arranged by special contract with those residents of the commune whose names are on the registered list already referred to, and who are approved by the permanent committee, as persons likely to give proper attention to the insane committed to their care. No house, unless by special privilege, is allowed to have, at one time, more than two insane patients. The law excludes the homicidal, suicidal, and incendiary, those requiring continual restraint, and others unfit for family care. If a case is decided by the committee to be unsuitable, the patient is returned to the committing party. About ten per cent of each year's arrivals are so returned. More women are usually received than men. The patients are chiefly from the provinces of Brabant, Anvers, and Liége. There are a few from foreign countries. It appears that the increasing numbers of the insane have created a pressure upon asylum accommodation throughout the whole of Belgium, causing a greater demand for cottage provision at Gheel.

At the time of my visit there were in the commune about 238 private boarders, or *pensionnaires*, as they are called, who were kept at prices arranged with the *hôte*. This is a term applied to those who keep paying patients; those having paupers are termed *nourriciers*. A few of these paying boarders were received at as high as 3,000 francs each per annum, about thirty were boarded at 1,200 francs each, and two hundred at 400 francs each. About forty of the private boarders were foreigners. The paupers numbering 1,400 were received by *nourriciers* at charges fixed annually by the

Government. There are three classes of pauper patients. The first, designated as "*ordinaires*," are those who are in fair physical health and who are able to make themselves useful to some extent; the second, the "*semi-gâteux*," are those who can do a little work and are at times uncleanly; the third, the "*gâteux*," are those who can do no work and are uncleanly in their habits. The rates payable per annum for each patient belonging to these different classes were, for the first class, 219 francs; for the second, 255 francs; and for the third, 313 francs.

A certain percentage of the money received for boarding patients is reserved by the Government to meet the expenses of medical care, supervision, medicines, and clothing, and for maintaining the infirmary. The tax is so adjusted as to amply meet the expenditures made for these purposes, and varies in different years. The sum reserved from the payment made for each pauper patient the year preceding my visit, for administration expenses, medicines, supervision, etc., was 9 centimes per day, and for clothing, 10 centimes, making a total of 19 centimes per day, or 69 francs, 35 centimes per annum. Medicines are prescribed by the doctors and supplied by the apothecaries of the colony at fixed rates.

Since my visit the prices paid by communes for the board of indigent patients have been advanced so that the per-capita rate, including governmental supervision, medical visitation, medicine, and clothing, is, for the "*ordinaires*," $1.09, for the "*semi-gâteux*," $1.22, and for the "*gâteux*," $1.43 each per week. The average cost of maintaining ordinary patients in asylums for the insane in Belgium is estimated at $1.55 per capita per week.

When a boarding-house keeper is unwilling to retain a patient on account of violence, or for any other cause, the matter is brought before the visiting inspector, or *garde de section*, who reports to the director, and an order is made for removal to the central station, until provision is made elsewhere for the reception of the case. Such removals are frequently effected, and it often happens that patients are transferred from one house to another in the colony.

VISITATION OF THE TOWN DWELLINGS.

The primary aim of the writer in visiting Gheel was to make an examination of the condition of the insane in their cottage homes. By courtesy of the chief physician, a guide was furnished who was, at the time, one of the district inspectors, and who had been for many years in the service of the institution. The succeeding days, which were spent in the company of this intelligent official, were wholly taken up in going from house to house, and from cottage to cottage, first through the town and afterward into the surrounding country. As we set out upon our tour, we met upon the street two well-dressed, middle-aged persons of gentlemanly appearance, who, after saluting us, held a few minutes' conversation with our guide. But for a casual remark from him, we probably should not have observed that they were insane.

It has already been shown that there is considerable difference in the charges for board. There is equal diversity in the character of the houses and their accommodation. Between the low, thatched house with mud walls, or the one-story building with attic-loft occupied by the pauper class, and the more inviting and more commodious two-story dwelling with plastered exterior, where reside the higher-paying patients, there are various grades of buildings. In the humbler dwellings of the town one generally enters the principal apartment of the house, which serves the various purposes of kitchen, sitting, dining, and reception room. In some of the houses the floors of these apartments are on a level with the street; in others they are even lower, and there were numerous indications of dampness. The fuel, of which less is required than in colder climates, consists here mainly of bituminous coal, which gives a clear bright fire. The water is obtained from wells. The streets have surface drainage. House slops are used to irrigate the garden.

Provided with a map of the commune, we proceeded to inspect some of the town dwellings. At the outset, permission was kindly given us to choose from the entire colony the houses and localities to be visited.

Our attention was first turned to one of the better class of the town houses—a goodly sized, convenient, two-story structure. Here resided two patients. One of these, an English architect, had gone out. In his room were evidences that he employed part of his time in drawing. Some studies from nature showed considerable skill in execution, as did also some sketches of a humorous character. The apartments were clean and comfortable. The sitting-room, which was used by both patients, was furnished with a sideboard, mirror, and a few pictures. On the centre of the floor was a rug. The adjoining dining-room, with sanded tile floor, also showed some taste, being hung with suitable pictures. Each of the two patients paid 1,200 francs a year and provided his own clothing. We have mentioned that the architect patient was out. His companion, a man of solitary habits, was still in bed, although the hour was ten o'clock. The architect's sleeping apartment, commanding a front view, had a wooden floor, and was furnished with fur rugs and ordinary bedroom furniture. The window and French bedstead were curtained. Here were all the needed comforts, as well as some of the elegancies of life.

On the opposite side of the street was a beer saloon, and

next to this, a spacious concert hall, in which a musical society held its meetings. Four or five of the members of this society were patients, one of whom was the leading violinist of the orchestra. The hall was fitted with galleries, and lighted with large chandeliers. The other furnishing comprised many small tables and numerous chairs. This concert-room opened rearward into a small pleasure garden. The usual entrance to the hall was through the beer saloon. The director of the musical society was also proprietor of the beer house, and one of the insane boarded with him paying 1,000 francs per annum. This patient made himself quite useful about the hall.

In the front part of the next two-story building visited was a small shop, and in the rear of it a parlor, in which were two boarders. At the head of a narrow steep staircase, closed at the top by a trap-door, was a small front bedroom. Here was a French bedstead of box pattern. The bed was good, but not very clean. There were also in the room a chair, wash-stand, and rug before the bed. The second bedroom, with an outlook to the rear, contained a bed, chest, wash-stand, stove, and garments hung on hooks. The foot-board of the bed had been broken by the violence of the patient. The house was small.

Passing to the adjoining cottage, we found a reception-room with a clean tile floor. The furnishing was plain but appropriate, and there were flowers in the window. Here was heard the homely tick of the old-fashioned clock on the kitchen wall. Altogether the accommodation was good, but the stairs were steeper and narrower even than those last mentioned—seeming to be positively dangerous. The bedroom contained a French wooden bedstead, with blankets clean and comfortable, and a feather pillow over another of wool. The window was of medium size and grated with bars about six inches apart. A wardrobe and a closet were in the room and a rug was before the bed. In the second sleeping-room the bed was good, the bare floor was clean, and the walls were white. The window was without grating. Both were back rooms, having outlooks upon a neatly kept garden and the country beyond. The patients here paid 600 francs each per annum. The whole house seemed tidily kept.

The floor of the kitchen in the next one-story cottage entered was sanded. Food was cooking in the old-fashioned fire-place. Every thing here was very plain. The only patient, an insane man, was preparing potatoes for dinner, the other members of the household being out in the fields. We were told that he was of an affectionate disposition, and that when a child of the family died a short time previous, his grief was inconsolable. The plates and furniture were clean, the kettles very bright, and the whole interior indicated good housewifery. Off the kitchen was a moderate-sized bedroom for the patient, who paid 255 francs per annum. The window of this room had wooden slats. The bedstead was of the box pattern, and the bedding sufficient.

The next house and the last mentioned were under the same roof, and their accommodation and housekeeping were similar. In the day-room or kitchen an epileptic patient sat in a corner behind a table, with a chair in front of her. The housewife was busy preparing cabbage for dinner in the same apartment. In the old-fashioned fire-place was a stove. The rafters showed overhead. In the adjoining room was an iron box-bedstead. The bedding, including two pillows, was clean and comfortable. The tile floor was sanded. The window, 2 x 3½ feet, had hinged sash, and was protected by three bars. In a corner, under a bench, was a heap of sand for use on the floor. There were two chairs here, and the clothing was hung on nails.

As already indicated, lace-making is one of the leading industries of Gheel. In some instances this is carried on in houses where patients board. An example of this came under notice in the next house visited, where three girls— not insane—were thus engaged. It was said that the female insane are sometimes serviceable in this occupation. One of two patients here, a young woman, was seen walking about outside. She complained that, as she spoke nothing but French, she could neither understand nor be understood by the household, all of whom were Flemish. Attractive in face and figure, and evidently of an erotic disposition, she did not seem suited to the large degree of liberty granted her. When in good humor, she was said to be an active worker. The second patient sat in the kitchen, and, it was said, did no work. Both were suitably dressed, but wore wooden shoes. The bedrooms were tidy, and without gratings or other security at the windows. One of the patients slept down-stairs, the other in an attic. The stairs were both narrow and steep.

The adjoining dwelling had better accommodations than the one last inspected. The housekeeping was of about the same standard. The mid-day meal had just been placed upon the table. There was a plate, cup, fork, and spoon for each member of the household. An iron dish in the centre of the table contained the food, which was served out with a ladle. Two patients sat waiting for dinner. One had a plate, spoon, and fork in her lap. It was stated that she made herself useful peeling potatoes. Her companion did no work. On the lower floor was the bedroom of one of the patients, which was of fair size, and contained a bedstead of box pattern with plenty of bedding. The tile floor was sanded, but it was not quite clean. The window had three bars. At one end of the attic was a small bedroom occupied by the other patient. Here was a box-pattern bedstead, and apparently sufficient bedding, though the surroundings were those of poverty. The stairs were dangerously steep.

The resident proprietor of the next house, who was not engaged in business, had returned to Gheel after a nine years' sojourn in America. His two-story building, with red tile roof, had been recently erected, and evidenced the improved fortune of its owner. Here were two private patients, paying respectively 1,400 francs and 1,000 francs per annum. One of the bedrooms was furnished with a French mahogany bedstead, very good bedding, rug, upholstered chairs, table, wash-stand, etc. The other bedroom, fully carpeted, contained a mahogany bureau and looking-glass. A commodious front room, well-furnished, was used as a sitting-room by the patients, as well as by the family. The interior woodwork of the house was painted; the walls were papered; and the apartments, including the bedrooms, were lighted by large windows without gratings or other restrictions. The whole interior was tidily kept, and the atmosphere was pure. In short, the surroundings of the patients were seemingly all that could be desired. The character of the housekeeping was indicated by snow-white curtains and spotless linen. One patient, who, it was said, gave no trouble, had just gone out to a neighboring inn for a glass of beer. He had formerly been a journalist in Switzerland. His companion, who was out walking in the back garden of the house, was said to be at times troublesome, giving many orders and causing annoyance. They breakfasted at 7 A.M., dined from 12 to 1 P.M., had coffee at 3 P.M., and supper at 7 P.M. The diet was varied, and the patients were humored in their choice of food as far as possible.

A large mansion, built close to the street, cannot fail to

arrest the attention of the visitor to Gheel. This was formerly the residence of Dr. Bulckens, through whose humanitarian efforts in the capacity of head director great reforms were accomplished throughout the colony, and whose lamented death occurred in 1876. The house was occupied by his daughter, who received two patients of the highest class. The broad central hall of this mansion, paved with black and white marble, opened into grounds at the rear in which were flowers, shade-trees, and convenient seats. The furnishings of the reception, sitting, and other rooms were in good taste. A large parlor on the ground-floor was assigned to one of the patients. The richly carved bedstead and other mahogany furniture were highly polished, and among the various ornamental objects was a vase containing a luxuriant ivy. The patient who used this apartment was, at the time of my visit, in the kitchen, where he was having a cup of coffee after dinner. The upper bedrooms and other portions of the house were furnished on a similar scale of comfort and elegance. One of the patients, an Englishman, was said to behave very badly at times, using improper language, and manifesting dissatisfaction by attacking the master of the house, though his violence never extended to children or to the hostess. We were informed that he rarely broke anything, never tried to run away, and never molested any of the servants. No means of restraint were resorted to, an effort being made to calm his excitement by kind treatment. In another apartment was a patient whose manners and appearance betokened a gentleman of culture. In his room was a collection of books which evinced fine literary taste. This patient was said to be a Polish prince, and this was his second sojourn here. He had previously left Gheel apparently cured, but, suffering a relapse, he returned and requested that he might be received into the same house.

Leaving apartments which were exceptional in style and completeness, we next visited an abode in striking contrast. Here two girls and a boy were at play in the living-room or kitchen, the floor of which was almost level with the street. An older boy sat, tailor fashion, on a table sewing. One woman was also busy plying the needle, while another was engaged at housework. There were two female patients here. The kitchen floor was laid with brick, and it was rather dirty. Some of the house furnishings showed signs of comfort. Climbing a very steep flight of stairs, we reached an attic chamber of large size, in which was a wooden box-pattern bedstead that had two straw mattresses, one above the other, clean linen, and two pillows. The small window had hinged sash, and was protected by four slender wooden bars. The door had a wooden latch and an outside bolt. The second bedroom, 9 x 10 feet, contained, like the first, a good bed, and the floor, like that of the larger apartment, was clean. The smaller room was lighted from a single pane in the roof. Here were signs of dampness, and there was no means of ventilation.

In a narrow alley leading from the main street, and terminating in a garden plot cultivated by those living in its vicinity, we came upon a low block of several dwellings. Entering one of these by the street door, we found ourselves in an apartment fulfilling the several offices of kitchen, sitting-room, and dining-room for a household consisting of a husband, wife, seven children, and two female patients. Here the head of the household was sitting by the fire with a child on his knee—the picture of domestic felicity. The husband and wife, together with five of the children, slept in a room scantily lighted from a window opening into the living-room of the house. This sleeping apartment was in a state of confusion, being filled with furniture and clothing. A steep flight of stairs led from this back room to the attic, where the two insane women and the two eldest children slept. The patients occupied respectively the front and back portions of this attic, while the children slept in an intermediate open space.

In the adjoining cottage of similar internal construction to that just mentioned resided a man and his wife, one child, and two female patients, one of whom, reputed to be idle and nervous, was in the habit of beating the mistress of the house. It was said that the next-door neighbor had sometimes to be called in to assist in restraining her when violent. The other patient was also indisposed to work. Here, as in many other of the humble abodes, the poverty of the mistress denied her the luxury of stockings. Her wooden shoes caused an awkward gait and made a clattering sound upon the floor in walking. In the living-room, or kitchen, colored prints of saints in rude frames were hung on the wall, upon which were also a crucifix, and a mirror surrounded by photographs. The husband, wife, and child slept in a dark bedroom back of the kitchen. The floor of the latter was sanded. The patients slept up-stairs. In one of their bedrooms was an iron bedstead, box-pattern, with temporary foot-board. The bed looked comfortable, having two pillows, clean linen, and a straw mattress, underneath which was a tick filled with straw. The floor was bare, and the furniture comprised one chair, a chest, and a short rack for clothes. The window in the roof had two panes, each about 9 x 16 inches. There was no means of ventilation. In the larger front attic room there was a French wooden bedstead with a straw bed, a wool mattress, and plenty of bed-clothing which was passably clean. The other furniture consisted of a chest and a rack for clothes.

The windows of the next house had clean white dimity curtains. The occupant, a widow, stated that she had five children away from home, and residing with her, one son, an aged father, and two female patients. One of the latter, a middle-aged woman, was sewing in the front room. The other, who was much older, and deaf and dumb, sat idly by the fire. She had been here for a great many years, and had helped the widow to rear and care for her large family. The proprietress, for this reason, held her in great esteem. In this patient's bedroom we were shown a little basket in a corner containing two rosy-cheeked dolls. These, it was said, received much of the patient's attention, being even taken to bed by her at night. We were also shown a box of toys and trifles used for the gratification of the imaginary children. The widow was evidently affected in speaking of these manifestations of a motherly instinct— apparently the only unbroken link obscurely connecting this poor creature with the world of intelligence. On the mantel-shelf of the day-room was the customary crucifix, and on the wall an old pendulum clock. Over the door leading to the back room was a figure of the Virgin, with two bright silver candlesticks, one on either side. In the attic, two rooms were partitioned off as bedrooms. Both of these had wooden bedsteads, clean and comfortable beds, chairs, and clothes-racks. Each was lighted by a large pane in the roof. There was no means of ventilation. One of the chamber doors was bolted. Here, as elsewhere in Gheel, the bolt was on the outside of the door of the patient's sleeping-room. The register-book showed that the doctor had made thirteen visits during the preceding twelve months.

Before leaving this alley we noticed at its extremity an out-house having a number of stalls without doors. It was for the common use of those in the neighborhood. It had a shallow open pit underneath, was in full view of the passers-by, and gave rise to gross violations of propriety, an instance of which came under our observation.

Entering another cottage, we found its inmates were the woman of the house, her grown-up daughter, and two patients—one an adult female, the other an idiot child. The elder patient had gone out in charge of the grandchild of the mistress. The idiot child was filthy and required much attention. The house, nevertheless, looked clean. The grown-up patient was boarded for 219 francs and the child for 313 francs per annum. The former slept in a small, low-roofed back room, having a small grated window. The bolt on the outside of the door, it was said, was never used. Bars and bolts were here spoken of as relics of the past.

In another large cottage resided a widow and three grown-up daughters. The occupants were all tidily dressed, the house was neatly kept, and the furnishing was good. There were here two patients, each paying 316 francs a year. One of these, an old woman, said to be of uncleanly habits, had a room on the ground-floor.

OUT IN THE COUNTRY.

The houses visited up to this point were in the town. Afterwards an examination was made of the cottages in the thickly settled surrounding parishes. In the inspection of this part of the colony a two-wheeled vehicle peculiar to this country, having seats for four persons, was used. The roads in some portions of the outskirts were found to be even worse than those in the prairie regions of America, being in wet seasons very muddy and almost impassable. Many of the houses indicated that the lives of the occupants were little else than a struggle for bare subsistence. The fuel used in the daytime for cooking was mainly brushwood. In the evening peat is burned.

The first farm cottage we examined was a one-story building with straw-thatched roof. In one end resided the family, and the other was used as a stable. The well was but a few feet from the door of the house, and about twenty feet from the stable. The patients were out, and the only person in the house was a woman who was working in the kitchen. She wore no stockings, and her feet were protected only by wooden shoes.

Leaving these ill-conditioned premises, we soon came to a low building with thatched roof, which, like the former, was rented. Here, also, one end was allotted to the family, and the other to the cattle. Close to the door of the house was the manure pile. The tile floor was level with the ground, and damp. The well was only ten feet from the house and stable. The mistress wore a dirty dress, clogs, and no stockings. There were two male patients in the kitchen, or living-room. One was idle, the other was peeling potatoes near the fire-place. This was ten feet broad, three feet deep, and there was a baking-oven on one side. As is usual in these farm cottages, the fire was made on the hearth. From a high wooden crane was suspended a large kettle used for boiling food for cattle. This was so arranged that the kettle and its contents could be swung round to a door opening from the kitchen into the stable.

Another mode in Flemish farm-houses of conveying cooked food to stock is by means of a semi-circular rail near the ceiling, extending from the kitchen fire-place to the stable door, there being thus direct internal communication between the culinary and cattle departments, though at some sacrifice of pure air.

The next farm cottage was owned by its occupant, and

was in a little better condition. It, too, had stable communication with the kitchen fire-place and a like arrangement for conveying the food to the stable. The floor was little higher than the ground outside. The house was occupied by a man and his wife, six children, and two patients, both of whom had just returned from field work on account of rain. The man of the house and his two boarders wore wooden shoes without stockings. The bedrooms were at opposite corners of one end of the house. The beds were comfortable, having double mattresses and plenty of bedclothes. The tile floors were sanded, and in the outer wall there was a small aperture for the escape of water used in washing the floor. Between the sleeping-rooms and the kitchen was the family room, in which there was a bed. On a line stretched across the apartment were hung clothes to dry. The family slept in the attic. In the kitchen was seen the familiar clock against the wall, also a baking-oven opening into the large fire-place. The fire, made of brushwood, was in a cresset. Kettles were suspended by a saw-bar, so that they could be adjusted at convenience. Attached to the trunk of an old tree was a wooden crane used for raising and lowering the buckets in the well, which was but a few feet from the door. The stable, which was under the same roof with the dwelling-house, contained a calf and a goat.

The next rented cottage visited was of similar construction to the last. The family, including a man, his wife, and five children, were at dinner in the principal room or kitchen. They were seated at a small square table in the centre of the apartment. The first part of the meal consisted of a thin soup or porridge of celery and rice, in which pork had been boiled. A plate of potatoes and boiled pork followed the soup. The porridge was palatable and the bread was good. Seated at a low side-table, near the door, were two male patients eating the food which had been served to them. They were allowed spoons and forks. This side-table was half covered with soiled garments, including trousers, overalls, old stockings, a shawl, jacket, etc. The patients were dressed in fustian, and wore caps, stockings, and wooden shoes. One worked in the fields, the other in the house. They were said to give no trouble, and were boarded for 219 francs each per annum. A door opened from the room in which the family were eating into the stable adjoining, and, as usual under this arrangement, the close proximity of the latter was disagreeably perceptible. The bedrooms were somewhat larger than usual, one measuring 7 x 12, the other 9 x 10 feet. The floors were of tile, and, in conformity with general custom, a hole through the outer wall permitted the escape of floor rinsings. The windows were small and had hinged sashes secured by iron bars. By night the doors were said to be strongly bolted on the outside. Each room had a chair and a clothes-rack. The bedding, though plentiful, was only moderately clean. There was no ventilation. Close to the house was a large heap of stable manure, and, at a distance of twelve feet from it was the well. All the family, children included, wore stockings and wooden shoes.

The rule requiring patients to take their meals at the same table with the *hôte* or *nourricier* was said to be difficult to enforce. A deviation from the regulation, as noticed at this house, met the strong disapproval of the inspector, who made note of it in his official journal.

The next farm-house examined was, in point of order and cleanliness, an improvement on those that have been described. Its proprietor worked twenty-five acres of land, one half of which he rented. One fifth was in pasture, the rest in cultivated fields. His stock comprised eight cows, two calves, and three pigs. The kitchen or main room was large. The beams overhead were quaintly carved. There were rows of metal dishes in racks above the fire-place, and all metal articles were brightly polished. The fire-place did not differ from those already described. The inmates were a man and his wife, seven children, and one male patient, who was said to be very good, and who was received at 219 francs a year. He appeared strong and healthy, was attired in cloth cap, fustian clothes, and stockings, and wore wooden shoes. He did little work beyond peeling potatoes, for which they said they gave him at times a little money to spend on Sunday. In the sitting-room there were pictures on the walls and canary-birds singing in their cages. The furniture included a high antique clock, an old-fashioned dresser, and a large wardrobe. A lamp placed in a small one-paned window in the wall, between the sitting-room and the stable, gave light to both places. A long rail and pulley overhead were here used, instead of a crane, for moving the large kettle from the kitchen fire into the adjoining stable. The patient's bedroom, on the same floor, measuring 9 x 12 feet, had a comfortable curtained bed, and a large two-sash hinged window without grating. Underneath the window was the usual outlet for water used in cleaning the floor. The door was without bolts, and the furnishings included a chair and clothes-rack. The mid-day meal was preparing. It consisted of soup and pork, with white and brown bread. In the stable, close by the kitchen, were two cows.

Passing on to another cottage also occupied by its owner, we saw, as in some other places visited, that the hedges were well kept, and that in front of the house there was a plentifully stocked kitchen garden. This dwelling was larger than some of those previously visited, and tolerably clean-looking. The main room, or kitchen, had a low roof. There was a large fire-place here and an iron rail arrangement for conveying food from the kitchen fire to the stable, in which were several cows and a goat. In the outer wall of the kitchen there was an aperture for letting out the floor rinsings. In a separate sitting-room the walls showed signs of dampness, being discolored to the extent of nearly three feet above the ground. The mistress of the house wore stockings with her wooden shoes. The only patient, an old man, was seated in a corner by the fire-place. He had on wooden shoes, but no stockings. We were informed that he had been here nearly half a century. His sleeping-room measured 7 x 8 feet, and was 6½ feet high. The furniture consisted of a wooden bedstead and a chair. The patient's clothes were hung on wooden pegs. The bedding, including a straw bed and an upper mattress, appeared clean and comfortable. The floor was of tile and sanded. The window, measuring 20 x 30 inches, had one sash and no grating. There was no bolt on the door. The walls were tinted. The register kept here was the oldest we had yet seen, showing systematic medical visitation extending back to the year 1825.

Another farm dwelling inspected was also occupied by its proprietor. The kitchen was smaller than those just described. The floor was only a few inches higher than the ground outside. The door of the living-room opened into the stable. On the walls were several photographs, pictures of saints, and a crucifix. A quaint old clock, convenient clothes-presses, and a lively canary were among the other noticeable objects here. The beams overhead were discolored with smoke. Two female patients approached as we were taking notes. They were plainly dressed and wore wooden shoes. One muttering to herself entered the house and walked out again. Both seemed excited, talking loudly and violently gesticulating. In the corner of the kitchen stood the manure-fork and a broom. A side room was used both as a sitting and sleeping room. The bedrooms had tile floors, windows 20 x 30 inches, and each door had two bars and a bolt on the outside. In the walls were holes for the escape of floor rinsings. The beds and bedding were comfortable. The house was, on the whole, passably clean. We were told that during the day the patients talked incessantly, but that they were quiet during the night. Neither of them worked, and belonging to the ordinary class of patients, they were received at 219 francs each per annum.

Leaving the foregoing farm cottages, which are situated far out in the country, we approached a small hamlet, where the spectacle of a vehicle containing two unannounced strangers in the company of a *garde de section*, appeared to be an event so unusual as to awaken considerable interest among the juvenile portion of this secluded community. As we paused to survey a one-and-a-half-story brick building before entering it, a wondering, interested, and barefooted little crowd gathered about us at the door. It consisted of seven little girls and ten little boys. At a more respectful distance stood several adults, among whom was a harmless idiot.

Entering the roomy kitchen, we found ourselves in what served the double purpose of beer-house and dwelling-house. The fire-place measured about fourteen feet across, or nearly the whole breadth of the kitchen, which forms, as usual, the principal apartment in these houses. A small kitchen grate was in one corner of the fire-place, and in the centre of it, a cresset for a large kettle. A door opened from this room into the stable. On a long shelf were brass candlesticks, a small crucifix, and a row of crockery plates, which were kept in position against the chimney by a string. A carved beam overhead, with joists on either side of the fire-place, supported the roof. The stable emitted a strong odor. Only six feet from the house was the family well. Under the eaves' spout was a vat for rain-water. A patient whom we saw walking in the garden paid 400 francs per annum for his board. Ascending a very steep and narrow staircase terminating at a trap-door, we reached the bedroom, measuring 10 x 12 feet. It had a fair sized window without grating, a bedstead and a good bed, a rack and shelf for clothes, and one chair. There was neither carpet nor rug on the floor.

Before reaching another cottage we passed on our way several small, poor-looking houses, three sides of which were constructed of loam and stubble, and one side of brick. The inspector stated that the patients' sleeping-rooms were on the brick-built side. Up to this point we had occasionally seen patients at work with their guardians in the fields. The implements of husbandry seemed rude and primitive.

The next farm cottage visited was of one story, and the tenant rented along with it nearly nine English acres of land. It had a low thatched roof, brick walls, and was altogether very ancient-looking. A patient who did no work was standing outside. He wore woollen stockings and wooden clogs. The manure heap was only twenty feet from the kitchen door. The floor of the kitchen was but very little higher than the ground outside. This apartment had a large black fire-place, the smoke from which, after making futile efforts to escape, filled the room. The sooty rafters underlying the thatched roof showed overhead. By the fireside sat an excited and talkative female patient of advanced years. The chairs, repaired with straw and rope, were one of many indications of pinched living. Altogether, the condition was lower than in any of the cottages hitherto visited. The inmates were a man and his wife, six children, and two patients. It being an afternoon when there was no school, three girls were at home. They were all barefooted. One of the bedrooms measured 7 x 10 feet, and had a small barred window. The bed was not so good as the one we had last seen, nor could it be commended for cleanliness. It was 2 P.M., and the patient had just risen. The second bedroom was much the same in appearance, but rather cleaner.

In the next cottage dwelt its proprietor, who cultivated thirty-two acres of his own land. The kitchen garden was large. In the stable, which connected with the house and formed a part of it, were cows and calves, a yoke of working oxen, and several pigs. About forty feet equidistant from the dwelling and the well was a closet without a door, and two thatched and ruinous-looking structures used as a shelter for some portion of the farm stock, of which there was more than was usually seen on these small holdings. A vat of dirty water stood about ten feet from the house, and only eight feet from the well which was about the same distance from the door of the dwelling. The wellwater presented a green and impure appearance. Over the well was a wooden pole fastened to the trunk of a tree and used for raising and lowering the buckets. The tile floor of the cottage was lower than the ground outside, the entrance was dirty, and the interior wet with slops. The door between the kitchen and the stable was wide open, and the low beams overhead in the kitchen were black with smoke. On the table were the remnants of the previous meal, a quantity of salt, and a broken and dirty old comb. There were also in the room two chairs, a bench without a back, and some dirty plates and cooking utensils scattered about. The appearance of all the inmates indicated great personal neglect. The husband, wife, and two patients wore wooden shoes and dirty stockings, and the daughter was barefooted. The sleeping-rooms were up-stairs. One had a small grated window, and the box-bedstead had unbarked sticks across the bottom. The bed, insufficiently supplied with clothing, had not been made up, and was partly wet. The other room seemed a little less uncomfortable; but the bedstead was smaller, the floor bare, and an old broken chair without a seat was the only additional furniture. Its small window was protected by two bars. Access to these rooms was through an attic filled with rags, old furniture, and miscellaneous articles lying about in indescribable confusion. Our hostess good-naturedly conducted us through her establishment, seemingly well satisfied with her management and flattered by our visit. One of the patients was described as a hard worker; the only work the other did was to peel potatoes. These wretched quarters showed a scene of uncleanness, disorder, and discomfort, within as well as without, which left a most unfavorable impression upon the mind. "This," said our honest guide, "is the dirtiest place in the colony."

The last of the farm dwellings visited was occupied by its proprietor, and it had attached to it seven and a half acres of land. The thin walls of this cottage were of loam and stubble, and the roof was thatched. At one end was the stable, at the other the family abode. The floor was but four inches above the level of the ground outside, and in the wall were visible fractures. In the small apartment, which served as kitchen and living-room, the floor was of tile, the ceiling low and black, and the scanty furniture included a rickety table. The housewife wore wooden shoes, and she was begrimed with dirt. Here were four children, and one epileptic patient who did no work and seemed disturbed by our visit. The patient's bedroom, 10 x 12 feet, had a small window and tile floor. There were two bars and a bolt on the outside of the door. It was entered through another sleeping-room, from which it had been partitioned off. This had an earth floor. There were but few of the necessaries and an almost entire lack of the comforts and conveniences of life.

CLOSE OF INSPECTION.

My third day at Gheel was now drawing to a close, and from the number of houses visited, and the widely different circumstances in which I found the insane, the conclusion was reached that I had seen enough to form a correct opinion as to the care and treatment bestowed upon them throughout the colony.

As bearing on the general welfare of the insane in family care, it may be further stated, that those boarded with *hôte* or *nourricier* may be assigned work suitable to their capacity or inclination. This permission the authorities withdraw if they find that a patient has been overworked. Such matters come under the cognizance of the *gardes des sections*, whose visits are made at unexpected times. As an inducement to work, money rewards are sometimes given, and we were also informed that tractable and docile patients have extended to them desirable privileges, such as attending church, fairs, or festivals in company with the family.

Neither *hôte* nor *nourricier* is allowed to use mechanical appliances for restraint without the sanction of the sectional physician. As accounting for the presence of bolts and bars in so many of the houses, it should be stated that the care-takers are by law held responsible for any damage a patient may do, and for his capture in the event of escape. It is claimed that there are few acts of violence and that suicides are extremely rare. A tragic story is told of a burgomaster of the town who many years ago met his death at the hands of a patient. The number of escapes during the year preceding my visit was given as twelve, a total which had not been exceeded during any one of many previous years. In these cases the runaways were captured and returned, as might be expected in a community where the householders act in such emergencies as a body of police, and render one another assistance.

THE COLONY; ITS ADVANTAGES AND DISADVANTAGES.

The advantages of the Gheel system are so generally understood, that it is unnecessary to detail them here at any length. The natural conditions of home life where the patient is made one of the family circle participating in its religious privileges and social enjoyments, the occupation of mind in farm pursuits as well as indoor industries, the pecuniary recognition of labor, and the general freedom from restraint, may be enumerated as among the incalculable and unquestionable benefits of the colony. On the other hand, there are some serious defects, which cannot in any just estimate be lightly passed over. These are not so manifest on the surface, for the simple reason that there is a common interest to keep in the background evils which are inherent in the system and beyond the control of individual effort.

In considering the Gheel colony as a whole, the fact must never be lost sight of, that the insane admitted into the commune are a selected class and are boarded out under what may be termed a double recommendation or certificate. In the first place, the committing authorities regard them as suitable cases for family care, and, again, their fitness is, as it were, confirmed, on their passage through the observation station. Yet, as has been shown, many even of this selected class are necessarily returned, no less than ten per cent of each year's arrivals at the central station being rejected as unsuitable. The Gheel colony, therefore, does not present a comprehensive system adapted to all classes of the insane. It was further authoritatively stated that there were at Gheel few curable patients, most of them having undergone treatment before reaching the colony.

The salaries of the sectional physicians are small and their districts large. The sixteen hundred patients of the colony must be visited once a month and acute cases as often as their exigencies require. As this work is in addition to a private practice of the sectional physicians, its satisfactory performance must be difficult. In the inspection of the cottages, the register-books were examined, and it was found that, although there was the requisite number of medical visits, averaging one a month during the year, there were occasionally long intervals between visits. In one instance there was no entry made for ten weeks.

The number and proximity of places where intoxicating drinks are sold indicated a danger, though it must be admitted that no case of actual excess came under my observation. The law forbids, under penalty of fine or imprisonment, the sale of brandy or other spirits to the insane, yet these liquors are generally kept for sale in the public houses. Wines are too costly to come within the reach of any large number of the patients. But the only restriction on the sale of beer is that the insane must not be allowed to drink "too much." This may be a knotty point to determine when the definition is left to the beer-seller, who must make out of his business the maximum of profit consistent with the bare observance of rules.

The care of the insane is relied upon by the people of the place as their main business or means of money-making. It is therefore liable to abuses common to all undertakings that are governed by considerations of profit and loss. Yet it may be urged that it is the interest of every trader to maintain a good reputation, and to avoid acts of cruelty or neglect, which would not only injure his own business, but damage that of the entire colony. In this way an *esprit de corps* is established, which, no doubt, tends to prevent abuses of a flagrant character. There are, however, faults for which the care-takers cannot justly be held responsible. Prices are made so low by the authorities that it is found impracticable to meet, in all respects, the standard regulations. For example, there is a rule which requires that patients shall be supplied with fresh meat every day; but keepers find it difficult to observe this requirement. Nevertheless, it seemed on the whole that the pauper insane fared as well, if not better, than their guardians. The advance made since my visit in the prices paid for board of this class it is hoped has enabled the *nourriciers* to improve their dietary.

Humble as are the accommodations of the insane at Gheel, it is fair to presume that they are equal, in most cases, to those of the homes whence they came. Generally, the treatment of the patients in the several houses appeared to be considerate, and marked by a degree of proficiency indicating an instinctive and possibly hereditary tact in dealing with them, though among the *nourriciers* were some manifestly lacking in qualifications which would be deemed essential in an asylum attendant.

Considering the large extent of territory and the number of the insane, it is hardly possible that the supervision can be so complete as to insure the due observance of the rules and regulations established by the government for their protection. The straining after economy, which regulates the rate of maintenance, as has been shown, affects the quality of the food, and also influences, prejudicially, the nature of the medical and other supervision.

There appeared to be a lack of intelligent provision for the needs of the patients in regard to bathing facilities, ventilation, pure water, and other requisites. The damp floors, the primitive methods of getting rid of floor rinsings, and the too frequent proximity of stables and manure heaps to the farm cottages, show a lamentable indifference to sanitary considerations.

Incidental to the system is the opportunity for exercising the tendency, natural in children and young persons, to trifle with the insane and idiotic. Chafing was not observed to such an extent as to produce any great amount of irritation in the patients, yet it was sufficiently manifested in the streets to convince me that the practice was one which might lead to injurious results.

In proportion to population the number of idiots belonging to Gheel who were not boarders appeared to be abnormally large. No statistics bearing on this subject were obtained, but it would be interesting to know whether this seeming disproportion is in any way attributable to the presence of the insane in such large numbers among the general population. This question assumes some importance when we bear in mind the opinions which have been expressed upon it by medical men. No less an authority than Esquirol has made the following forcible observation: "To the inmates of a madman's family the sight of those acts which are committed by the insane may be highly prejudicial. A woman who is *enceinte* and easily excited would run some risk by living constantly with a person who is mentally deranged, and the example to children and young ladies might become a predisposing cause of mental disease."

Looking at the commune in its moral aspect, one cannot help thinking that the shockingly immodest exhibitions which here and there meet the eye must have a baneful influence on the large number of children of both sexes growing up in their midst. Nor can one approve of the presence of some insane young women in the colony, who, from their erotic natures, seemed helplessly exposed under the circumstances in which they were placed. That liberty, so beneficial to many of the mentally afflicted, was, in their case at least, unwisely bestowed. It appears that illegitimate births do sometimes occur here among the insane. One authority estimated these as happening not oftener than once in five years ; another maintained that they were more frequent. Any system, under which such occurrences are at all possible, is greatly at fault.

Gheel has been dealt with at some length because of its many points of interest, and for the additional reason that its boarding-out system has presented itself in quite different aspects to those who have studied its workings. After a careful examination of the colony, the writer is forced to the conclusion that the Gheel system is of little practical value to America, except as demonstrating that a great amount of freedom is possible in the care of certain classes of the insane.

CHAPTER VII.

THE PROVINCIAL INSANE ASYLUM OF ALT-SCHERBITZ.

OF the many asylums for the insane in Europe, there is none more interesting than that of Alt-Scherbitz, one of the two public institutions that meet the requirements of the Province of Saxony in Prussia. Here has been wrought out a system in which are incorporated some of the best and most modern methods of caring for the insane in England, Scotland, France, and other countries.

This asylum is situated in a fertile valley about a mile and a half from Schkeuditz, which is its station on the railway line from Halle to Leipzig. The central buildings, which comprise the administration department, reception stations, observation stations, detention houses, hospital for the bodily sick, etc., are entirely separate from one another, having no corridor connection, and are situated in the midst of improved grounds lying on the left of the highway leading to Leipzig. These are unpretentious brick structures, with outer porches. On the right of the highway are the domestic and industrial departments, the superintendent's residence (formerly the Manor House), and scattered cottages for patients. Near by is the small hamlet of Alt-Scherbitz, lying within the asylum estate, which contains upwards of seven hundred English acres. This property cost the round sum of one million marks (about $240,000). Ten of the cottages in the hamlet are owned by the asylum, and are occupied by quiet patients. Others are used as residences for the workmen connected with the institution. Through the estate flows the river Elster, a small stream with picturesque borders. On its right bank and about the superintendent's residence and the cottages are ornamental grounds. On one of the terraces here is a plain monument, erected, as its inscription shows, by the people of the Province of Saxony, in memory of the humane and intelligent founder of the asylum, Professor John Maurice Koeppe. The object of the institution, originally planned for 450 patients, but subsequently enlarged so as to accommodate 600, is "the recovery of the apparently curable, also the reception and care of the incurable and dangerous insane of the province."

Formerly all the Prussian asylums for the insane were under state control. Since the direction of local affairs has been intrusted to the provinces, they have managed their own institutions for the insane, deaf and dumb, and blind. The state, however, still retains its right of supervision over these institutions. The chief officer of the provincial government is the "Landes Director," who, with several councillors (Landes Rathen), directs the current affairs of the provincial government. For the disposal of matters beyond the jurisdiction of the "Landes Director" there is a so-called Provincial Committee, which meets about every two months, or as necessity demands. It consists of fourteen members of the Provincial Assembly. The latter body is composed of over one hundred delegates from all parts of the province, who convene once in two years to consider and transact the business of the province. With it rests the chief control of the institution and the appointment of the medical director or superintendent of the asylum. Reports of the medical work

EXPLANATION.

1 *Central Institution* { a *Administration Building,* b *Observation Station,*
c *Reception Station,* d *Internat H*
e *Hospital,* f *Mortuary.*
2 *Kitchen*
3 *Kitchen dwellings*
4 *Laundry dwellings*
5 *Laundry*
6 *Dairy with dwellings*
7 *Horse stable and dwellings*
8 *The Director's Residence*
9 *Farm-buildings*
10 *Assembly-house*
11 *Brickmaking establishment*
12 *Village houses, formerly belonging to villagers*
13 *Men's Villas*
14 *Women's Villas*
15 *Bath-houses*
16 *Pavilions for infirm chronic cases*
17 *Residence of one of the Assistant Physicians*

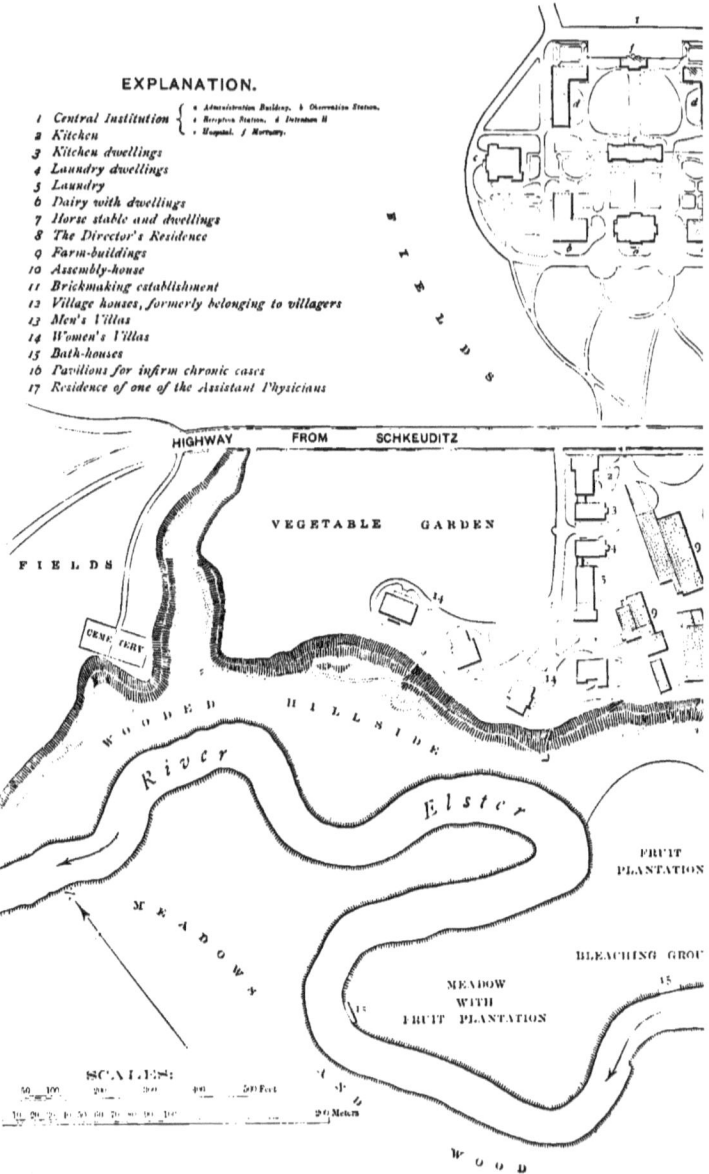

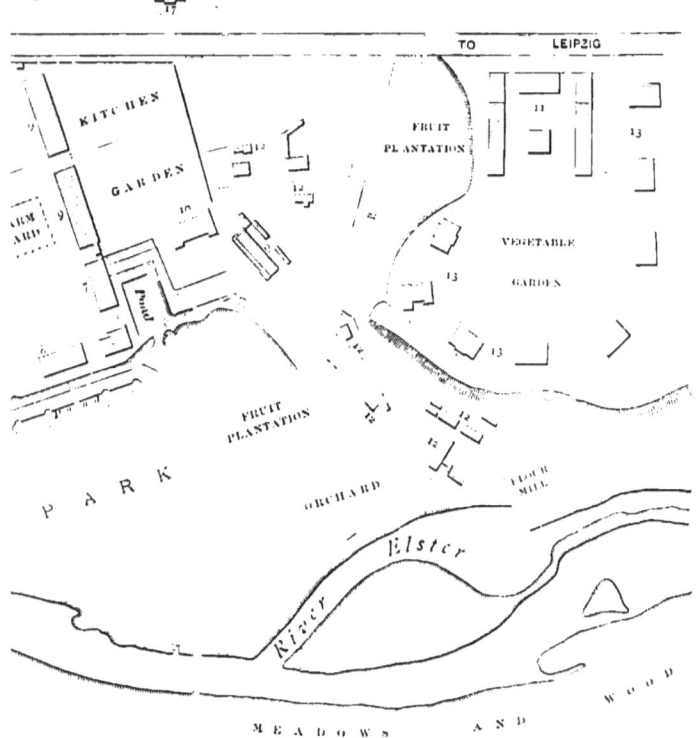

and of the administration are made to the chief of the provincial government, the " Landes Director," who resides at Merseburg ; likewise statistical reports of the entire result of the management to the statistical central bureau at Berlin. Under the supervision and inspection of the Provincial Committee and the " Landes Director " the affairs of the asylum are administered by its Superintendent, Dr. Paetz, who is appointed for life with the same right to a pension as other officers holding life appointments. He is assisted by a second, third, fourth, and fifth physician ; also by an accountant, a secretary, an assistant secretary, a steward, a house overseer, and a land overseer. His jurisdiction covers all the medical and economical affairs of the asylum and grounds or land. As the chief officer, he administers to subordinates the oath of office, and preserves order and discipline according to rules. He may be absent from the institution for three days without permission, if he arranges to have his place properly filled *ad interim;* and he may grant to any of his subordinates eight days' leave of absence.

The asylum must receive all curable patients, and of the incurable those who, at the same time, are dangerous or burdensome to their families or the public. Incurable patients not dangerous have also been admitted, and will be while the province has sufficient room in its asylums. Refusal to receive patients scarcely ever occurs.

Applications for admission, to which a certificate from the district physician is attached, are sent from the authorities where the patient resides to the asylum superintendent. The latter has the right in all urgent cases, also in all cases of curable or dangerous patients, to admit without delay. He communicates the proper intelligence to the local authorities and to the friends or relatives of the patient, and makes a short report thereon to the "Landes Director," which accompanies the application for admission. In cases not pressing, the asylum superintendent transmits the application for admission received by him to the "Landes Director," with a short discretionary opinion on what grounds the application was accepted or rejected, and, according to this submitted opinion, then the "Landes Director" will report to the local authorities who made application for the patient.

The insane are conveyed to the asylum with accompanying documents setting forth, among other things, whether the friends legally responsible are willing and able to bear the cost of the needed care in whole or in part, or whether this is to be defrayed out of means belonging to the patient. When there are particulars which, in the interest of the patient, his relatives, or friends, should be kept secret, but which are necessary to a perfect understanding of the case by the physicians, such particulars are excluded from the admission papers, and are sent separate, in a sealed envelope, to the superintendent of the asylum. The rules forbid the reception of an insane person who has been exposed to contagious or infectious diseases until six full weeks have elapsed after exposure. Upon the approval of the superintendent and under his regulations, patients are allowed to have special attendants.

After making arrangements with friends of the patient or others for his support, the superintendent, acting independently, makes absolute discharge. When dangerous patients, committed directly by the local authorities, have recovered, the latter are informed. Discharge on trial is allowed, and in all such cases the relatives, or authorities to which the insane are chargeable, undertake supervision, and every three months communicate with the asylum superintendent respecting the health of the patient. All such

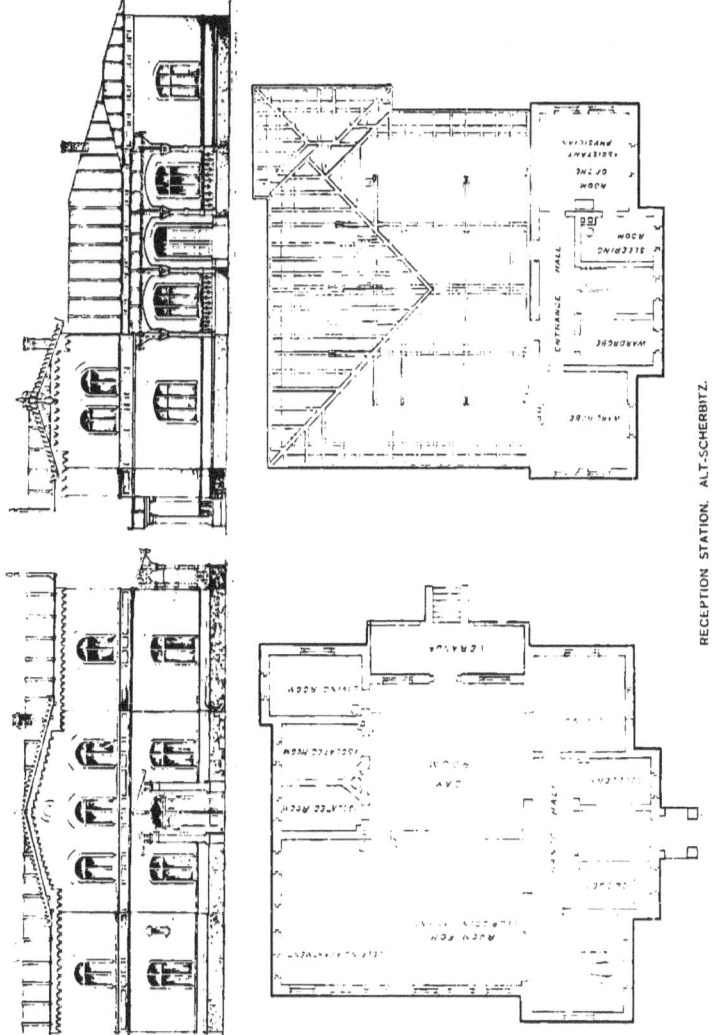

RECEPTION STATION. ALT.-SCHERBITZ.

reports are signed, not only by the relative or guardian, but by a medical authority. After two favorable reports, the name of the patient is struck off the books of the asylum.

The institution was opened in 1876, when forty insane persons were conveyed to the old farm-house of the Alt-Scherbitz manor. At the same time the central institution buildings were begun.

Patients of what is termed the " first class " were admitted at a yearly payment of 1,200 marks ($288); patients of the "second class" at 600 marks ($144); and patients of the "third class" at 240 marks ($57.60), for which sum the last named are provided with clothing in addition to maintenance. Any difference between the payment and the actual cost is defrayed by the provincial government. An additional charge is made for patients received from other provinces.

The two reception stations of the central institution—one for men and one for women—are clinical passage ways for all newly received patients, who are detained here as long as they need continual care and treatment. Those who remain in these stations are usually the acute curable cases, the incurable ones being soon transferred to other departments.

The observation stations—one for each sex—are for patients who, though not of the acute class, need special observation because they are not sufficiently capable of self-control nor reliable enough for reception in the colonial stations.

The two detention houses are for such male and female patients as it is necessary to restrict because of their being restless or dangerous, or from the suspicion of their having a desire to escape.

Centrally situated in the front of the group constituting the central institution is the administration building, the houses occupied by women being on the right side and those by men on the left. Rearward from this is placed the hospital for the bodily sick, having a department for each sex. Near the latter and completely hidden by trees is the mortuary.

The low brick walls which inclose the yards at the rear of the detention houses have sunken panels, and are surmounted at intervals by pillars supporting an architrave. They are architecturally designed to conceal their purpose, and are screened by shrubbery and creeping vines that twine gracefully about the pillars. Except in the isolation rooms, the windows of all the buildings belonging to the central asylum group have neither bars nor gratings.

On the first floor of the administration building is a reception room, a conference room, the superintendent's office, an accountant's office, a treasurer's office, and a porter's rooms. On the second floor are the living and other apartments of the second physician and those of the accountant.

The reception stations, built to accommodate fifteen patients each, are plain but comfortable buildings, and within and without resemble private dwellings. They have unlocked doors, and there is no suggestion of irksome restraint. In the building for women resides the head female attendant; and in that for men, one of the assistant physicians.

In the observation stations, each of which is intended to accommodate thirty-five patients, the furnishing was ample and comfortable. Here, as in some other departments, provision is made for open fires. The windows were, with one exception, unlocked, the exception being one with a lockable sash in a room that was less under the attendant's eye than the others. Windows of rooms that are opposite

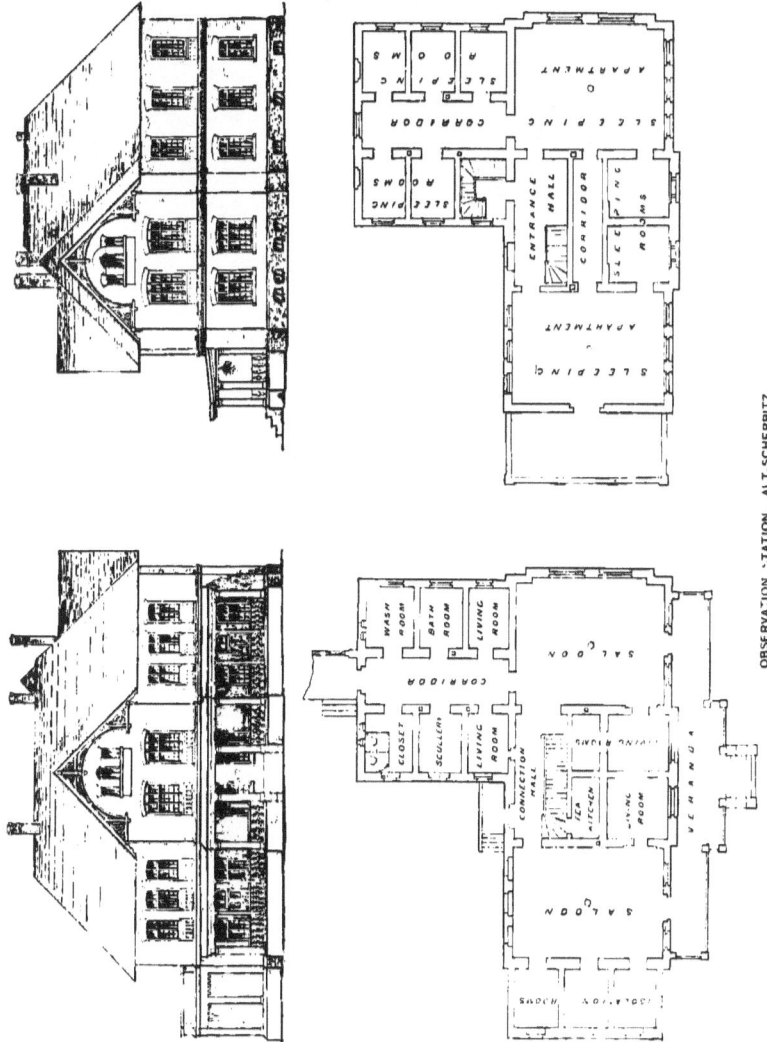

OBSERVATION STATION. ALT-SCHERBITZ.

and are occupied respectively by men and women, have frosted panes. The doors open rearward to shrubbery-planted plots, and in front to wide porches and the open grounds. Fine views, immediate and distant, are had from both windows and porches. In one of the rooms on the women's side there were two curtained couches for the use of such patients as desired occasional rest. In the windows and in other parts of the rooms were bright flowers. There is also an abundance of health-giving light. The sitting-rooms had plain chairs and sofas. Each station has separate sitting and dining rooms. Six single rooms are set apart for patients of a higher class. The sleeping apartments in the observation stations are mostly on the associate plan and were provided with iron bedsteads, each having a temporary foot-board, mattress of straw, India fibre, or horse-hair, over which was placed a second mattress, blanket, blanket cover, bolster, and pillow. The oiled wooden floors were uncarpeted.

The detention houses, like the other buildings, have numerous windows, from which there are pleasant outlooks. The interiors are cheerful and are comfortably furnished. Nearly all of the limited number of patients in the detention house for women were engaged in sewing or knitting, with an attendant in their midst. As we entered, a request was politely made to abstain from note-taking in the presence of the patients, it being explained that pains were taken to occupy their minds, and every thing likely to prove a cause of disturbance was avoided.

There are fourteen isolation rooms, which number is here considered more than sufficient for an asylum having six hundred patients. They generally stand empty in the daytime, and at night are used for sleeping-rooms for restless patients. In those used for isolating maniacal cases there are inside blinds to the windows. The isolation apartments are so distributed that there is in each reception station two and in each detention department five. There were formerly three rooms for isolation in each observation station, but these have not been used in a number of years for this purpose. The walls shutting off the grounds from the observation court having been found unnecessary, were removed.

In the management the tendency is towards an extension of the liberty of the patient. Isolation cannot be enforced without the consent of the Superintendent, who said: "It is avoided in every possible case." There were no cribs in use, nor were there any padded rooms. Patients who show a disposition to tear off their clothing are attired in strong garments buttoned at the back. This dress Dr. Paetz does not regard as a means of restraint. He says: "It does not hinder the patients from having free use of all their members, and this arrangement of the clothing only prevents its being torn off." He further says: "Every sort of restraint by force is strictly interdicted as being against the fundamental principles of the asylum. The patients enjoy the largest imaginable freedom, the asylum representing the non-restraint system in its widest sense. Restraint is easy to dispense with if one earnestly wishes to dispense with it." The number of nurses or attendants averages about one to ten patients. They live with the patients and lodge in the same dormitories.

The medical department appeared to be thoroughly systematized. Every patient on arrival at the institution must remain in bed on the morning after reception to wait a mental and bodily inspection by the whole of the medical staff. Each physician is required to write out separately his diagnosis of the case, giving his views of the necessary treatment. These become matters of record.

DETENTION HOUSE. ALT-SCHERBITZ

Patients who do not occupy the central institution are provided for in villas and other dwellings. Those for women are widely separated from those for men. The superintendent's residence and grounds, the asylum kitchen, dairy, laundry, and farm buildings are between. An assistant physician resides in one of the cottages belonging to the group occupied by male patients. There are three classes of cottages, or villas. The first and second classes are two-story buildings, and are all well furnished, the quality of furniture varying according to the payment for support. These buildings stand at a distance of about one hundred yards apart. From the most of them glimpses are obtained of the river Elster. They are separated from each other by neat low hedges, and attached to some of them are little gardens. There are no guards to any of the windows. The freedom of the place is shown by open doors, which everywhere meet the eye. It was a pleasant summer day when I was there, and the patients were passing in and out without interference, all, however, being under watchful supervision.

One of the large sitting or day rooms entered contained a circular sofa in the centre and an ordinary sofa against the wall. On the floor were some bright rugs, and the presence of flowers and objects of ornamentation lent an air of refinement to the whole interior. The other furnishing included mirrors, pianos, wardrobe, writing-desk, sewing tables, divans, and chests of drawers. Lamps were suspended from the neatly frescoed ceiling. The walls were colored. There are convenient store-rooms and closets in the several departments. The cottages, all of which are constructed to admit an abundance of sunlight, rest on dry, substantial foundations.

A third-class two-story cottage accommodating nineteen inmates had fewer conveniences than the others, though it was adequately furnished. On the lower floor is the sitting or work room, also the dining-room; and on the upper floor are the sleeping apartments. In the cottages[1] as also in the central institution[1] the arrangements for men are a counterpart of those for women. It is the aim of the management to have every thing relating to the care and treatment of the patients conform as nearly as practicable to home life.

The institution cottages in the adjacent hamlet of Alt-Scherbitz were next visited. At one of these the patient had gone out and locked his door, and the physician would not enter the dwelling without his permission. These buff-colored cottages with white mouldings present a neat exterior. On the doors are the names of the patients put on by themselves. One, manifesting some classical taste, had carefully imprinted in large characters over the door of his little cot the word "Salve." The furnishing of these houses is simple, yet sufficient for the needs of the inmates.

Provision for sixty harmless and infirm chronic insane of each sex is made in two cottage pavilions. One is on the right and the other on the left of a dwelling for an assistant physician and others connected with this department. These buildings form a separate group in the neighborhood of the central institution. They are two-story structures having associated dormitories on the upper floor.

There is one general kitchen for the whole establishment. The food is conveyed from it to the several departments. The vehicle containing the mid-day meal, and constructed so as to keep the contents warm, was seen on its

---

[1] While the plans and illustrations given of some of the Alt-Scherbitz buildings serve the valuable purpose of demonstrating the feasibility of substituting inexpensive, comfortable structures, something like ordinary dwellings, for the generally prevailing massive palatial edifices built on the congregate plan for the insane, they are not presented as faultless models for asylums in the United States.

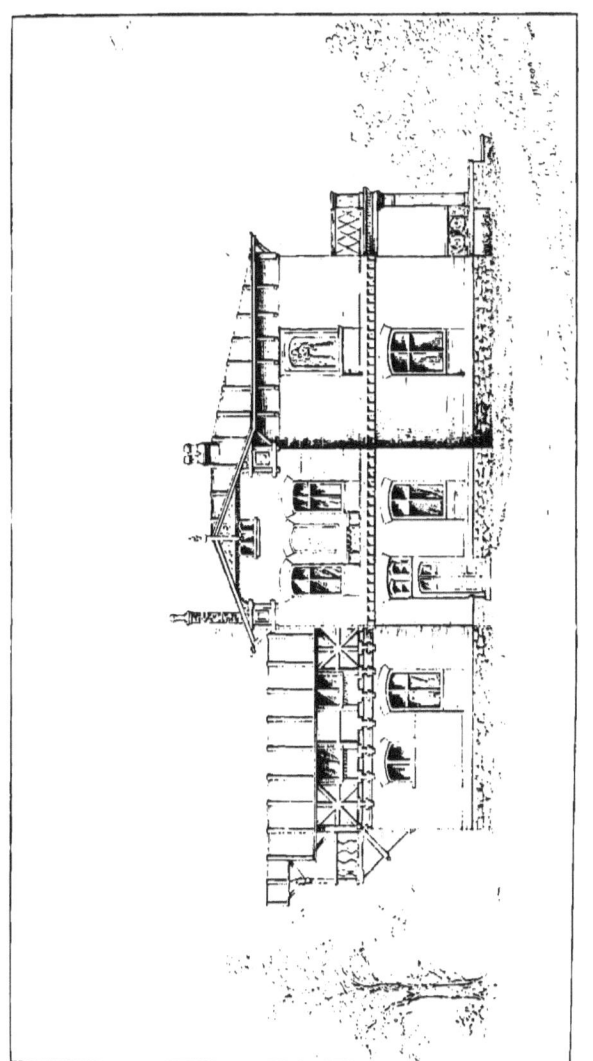

A VILLA FOR PATIENTS. ALT-SCHERBITZ.

way from the kitchen to the cottages. Two patients were in charge of it, one acting as driver and the other occupying a seat behind. In the extensive culinary department twelve patients were seen at work under the supervision of an attendant. Besides these general arrangements for the preparation of food, it should be mentioned that each domicile has its scullery and pantry. The rule as to food specifies that it be wholesome and well prepared, and further, that the dishes and cooking utensils be scrupulously clean.

The central department is warmed by hot air; in the isolating apartments the heating is effected by hot-water pipes; in the villas porcelain stoves are generally used, open fires supplementing, to some extent, the system of heating, and aiding ventilation. Water is elevated by a force-pump and distributed by gravitation. The bucket system is adopted, and all waste is utilized on the farm.

The farm buildings are extensive, and a large herd of cows is kept. The dairy, quite an important adjunct, was shown by those in charge with justifiable pride. There is a brickyard on the place, and from this source has not only been supplied brick used in the erection of asylum buildings, but a considerable quantity has been marketed. Other objects of interest which arrest the eye are a capacious hot-house, a well-stocked fish-pond, also an orchard in which are cherry, pear, apple, and plum trees.

Agreeable employment, suited to the mental and bodily condition of the patient, with relaxation in the form of outdoor and indoor games, is strongly enjoined by the management as a means of cure, and is carried out in the domestic department, in the several workshops, in the garden, and in the fields. I was informed that from eighty-five to ninety per cent of the patients were on the list of the employed. The women find much to occupy their minds and energies in the kitchen, the wash-house, and the dairy. Many work in the sewing-room, where wearing apparel is made, and where articles for institution use are repaired. Some of them milk the cows and work in the garden or in the fields. Most of the men are occupied at agricultural work and at different trades. The few not engaged in any form of employment include acute cases under medical treatment, those physically incapacitated, also those extraordinarily excited. "There is," says Dr. Paetz, "no branch of agricultural work in which the patients are not useful, and their use of implements has not been attended by any accident." In the brick and tile yards only a few paid laborers in addition to the patients are required to do the work. This is also the case in the other departments. In addition to the insane at work on the farm and in the brick-yard, others were employed as joiners, masons, wagon-makers, blacksmiths, carpenters, smiths, shoe-makers, tailors, saddlers, book-binders, stone-masons, painters, basket-makers, and clerks.

In furtherance of their recovery all the curable insane within this institution have equal care and consideration. Excursions for the benefit of the health of the inmates are periodically made, and in the case of the poorer patients the asylum bears the cost. Theatrical entertainments, concerts, dances, bowling, and a variety of games, are among the varied means of recreation and amusement. Particular attention is paid to cleanliness of the person and of the clothing. The latter is frequently changed, and the airing of all apartments is not a matter of hap-hazard, but of rule. A spacious swimming-bath is provided for the inmates in the river Elster. The correspondence of patients with outsiders is regulated by the asylum superintendent. The statute provides that complaints of want of proper care and

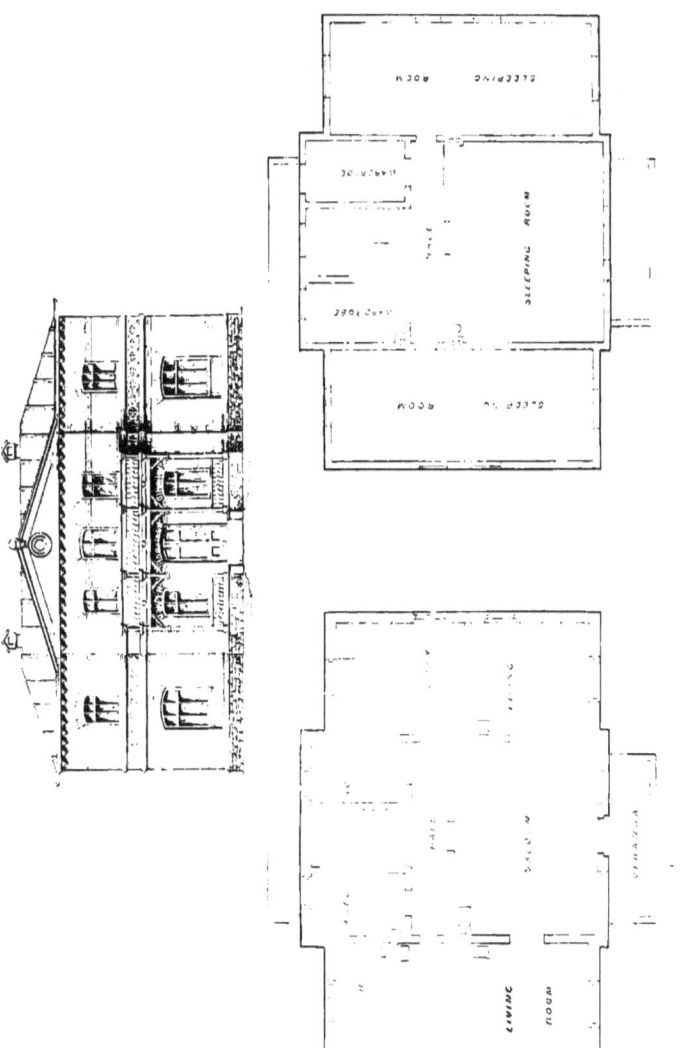

A VILLA FOR THIRD-CLASS PATIENTS, TWENTY-SIX BEDS, ALT-SCHERBITZ.

treatment may be made to the "Landes Director," and that charges of a grave nature may be brought to the attention of the Provincial Committee. Strangers are admitted to the institution only by permission of the superintendent, and he may deny the admission of a relative if he thinks the visit likely to be injurious to the patient. Care is taken to prevent annoyance to the inmates.

At this institution one is favorably impressed with the absence of barriers, the freedom from restraint, the kind treatment accorded the patients, the thorough supervision of the large numbers occupied in the various employments, the general atmosphere of cheerfulness and quiet, and the cleanliness and country-life aspect of the place. One instance illustrative of the intelligent management may be cited. In the laundry were observed a large number of women working at tubs with wash-boards. A few, however, stood idle and listless, with their untouched work before them. On inquiring of the physician why these patients were so placed while doing nothing, he replied: "They may be there for days and not raise a hand to work, but by and by the mind becomes interested, a slight effort is put forth, followed by another and yet another, until, at length, they become steady workers, and as a consequence their health is greatly improved." Such careful endeavor to engage the mind and induce volition on the part of the insane by kind and patient effort is a fundamental principle of the Alt-Scherbitz treatment.

Without any attempt at ambitious architectural display or costly interior furnishing, there is, apparently, in this institution every thing essential to the comfort of the insane. The whole system of care and treatment seems adapted to insure highly satisfactory results; and yet a stranger passing along the highway and catching glimpses of the asylum buildings through the trees and shrubbery, would hardly suspect from their unpretentious character and their arrangement upon the estate that the place was a public hospital and asylum for insane people.

CHAPTER VIII.

RÉSUMÉ.

A GENERAL survey of the field of inquiry to which the reader's attention has been directed in the foregoing pages suggests some thoughts which it may not be out of place to present, with a few conclusions drawn not only from this examination, but from other extended examinations of various methods of caring for the insane.

LOCATION.

It has been found in many instances that essentials to the good sanitary condition of an asylum and an economical administration of its affairs were overlooked in selecting the site ; and consequently the institution must suffer throughout its existence from mistakes that cannot be rectified.

The first consideration in the selection of a site should be its healthfulness. There should be an abundant, never-failing supply of pure water, opportunity for the ready disposal of sewage without endangering the health of the inmates or of the public by the pollution of streams or otherwise, a sufficient acreage of land, and easy communication by rail or water for both passengers and freight. The atmosphere should be pure and salubrious—free from malarial influences and artificial poisons. The situation should not be in proximity to any local nuisance or disturbing element. Although not always practicable to obtain without sacrificing more important advantages, it is desirable that water be supplied from a source sufficiently elevated to ensure its distribution throughout the various buildings by gravitation. The supply should equal at least sixty gallons a day for each inmate.

Many of the asylums formerly built for the insane were established within or adjacent to populous centres, where it was difficult to obtain sufficient land for gardening and farming purposes. This disadvantage led to a change of policy, and modern mixed asylums and those for the chronic insane are more generally located in the country and upon considerable tracts of land. Nevertheless, it seems to be almost universally the case that additional land is soon required, and it is frequently impossible to obtain it, even at an exorbitant price.

In order to overcome some of the difficulties arising from lack of sufficient landed estates in connection with institutions for the insane in England, the Government has made it practicable for one county to purchase land and establish an asylum in another county. It is often found necessary to adopt the unsatisfactory plan of leasing ground, sometimes at inconvenient distances from the asylum property. From past experience may be learned the lesson that it is always best, when founding an asylum, to secure at the outset sufficient land. If more should be purchased than is actually required, it could be sold at a profitable advance, as the improvements and high cultivation would increase its value. The area of land for an institution for the chronic insane should not be less than one acre per inmate. There are those who think half this amount is sufficient; but the weight of testimony is in favor of the larger acreage. Some American authorities place it as high as two and even three acres per inmate.

It is conceded that agreeable occupation, especially employment in the open air, greatly contributes to the health and contentment of the insane; and it has been demonstrated that the rate of maintaining the chronic insane is lowered as the quantity of land is increased; while in some instances, with a larger acreage, a higher dietary standard is attained. At Alt-Scherbitz in Prussia, Clermont in France, and the Willard Asylum in the State of New York, we find extensive vegetable gardens, goodly sized fruit orchards, wide fields yielding grain, and broad, rich meadows. The large herds of cattle kept at these places produce abundant milk, butter, cheese, and beef, for the consumption of the inmates.

Another advantage in having a liberal area of land is found in the opportunity afforded for reserving broad spaces for lawns and shady groves exclusively for the exercise and recreation of the patients, and for making a wider distribution of the various buildings, which is desirable in effecting a proper classification of the insane.

Not only is a large acreage of land desirable, but its quality is of great importance. That suited to gardening purposes seems best adapted for the uses of an insane asylum. The soil should be a warm fertile loam that may be cultivated early and late in the season for doing outdoor work— one that soon becomes dry after rain, and is thus suitable for recreation the greatest possible number of days in the year. It is much more agreeable to till good land than that which is poor, and the former will often return for the same amount of seed and labor double the harvest of the latter. Clay land and that which is stony should be rejected. Using the spade or plough on such, especially the latter, while trying the patience of a sane man, irritates the insane. A loamy soil by its absorbing qualities takes up impurities, thus acting in some degree as a disinfectant, and preserving a purer and more wholesome atmosphere. Further, when a large outlay in buildings is contemplated, it is unwise to place them on land other than that in every way suited to the purpose for which it is intended. The immediate site of all buildings should be on ground free from hidden springs, and capable of quick surface and effectual sub-drainage. A rocky substratum should be avoided, for it makes necessary excavations expensive.

The love of the beautiful in nature is common to mankind, and though it may not always be manifest, its peaceful influence is felt, in a greater or less degree, by the insane as well as the sane. Residents of the country, like those of the city, seek opportunities for recreation in the summer where nature is specially alluring, and this fact should not be lost sight of in selecting places for the care of the mentally diseased. A site commanding an extensive and varied prospect that tends to inspire pleasurable emotions in the mind, should therefore be sought. For a strictly curative institution—that is to say, for a hospital for the insane, where so large a tract of land as is requisite for an asylum is not necessary, nearness to a city or a large town is desirable.

In locating public charitable institutions, the State has not infrequently accepted donations of land made by local authorities or through the voluntary subscriptions of private citizens. It has invariably resulted that this temporary gain has proved a permanent disadvantage. By whatever name we call it, a gift of this character partakes of the nature of a bribe, and its acceptance creates the feeling that an equitable claim is thus established for the patronage of the institution; and if this be denied, ill-feeling and embarrassment to the management ensue. The result is more satisfactory when the State makes its selection on the merits of the site, irrespective of local gifts; and it is certainly more dignified to do so than to appear as a suppliant before a portion of its people. It would be well if every State should follow the example of Illinois, and forbid the free acceptance by commissions, of sites for public buildings.

BUILDINGS.

In making provision for the insane, we should not overlook the fact that there are two general classes or divisions —the acute and the chronic. Respecting the first, there exists, in a greater or less degree, the expectation of cure ; in regard to the second, the chance of recovery is reduced, it may be said, to a bare possibility. To the chronic class belongs the great mass of insane under care, variously estimated at from four fifths to seven eighths of the whole.

Every consideration of humanity demands that no time should be lost in applying remedial measures when they are most efficacious, and that, however elaborate the provision or expensive the treatment, no effort should be spared to cure the patient before the disease becomes chronic. Not only does every principle of humanity require this, but herein is true economy ; for, if an insane person can be cured before the acute period, which is comparatively brief, is passed, though the treatment be highly expensive, it is cheaper to do so than to maintain him during the following twelve or more years of his life in the chronic state, even if it be at a low rate of maintenance.

Past experience, it appears to me, has demonstrated that large mixed institutions are not effective agencies in the cure of insanity. In many particulars the requirements of the acute insane are more exacting than those of the chronic. These include special structural arrangements, very close medical attention, a large corps of specially qualified attendants, and a prescribed diet. If a standard of care suitable to the necessities of the acute insane is adopted by a large institution receiving both acute and chronic cases, it is unnecessarily expensive for the chronic insane; if the standard is made only conformable to the needs of the chronic, it is insufficient for the acute insane: and it has been found difficult to counteract the tendency towards a uniform standard of care for both classes under the same administration. Besides, a great number of patients increases the business responsibility of the superintendent and demands attention that should be given to the medical department, multiplies details, and finally individuality is lost, and the curative purpose of the institution, which should be paramount to every thing else, cannot be effectually carried out.

It is the opinion of the Hon. Francis Scott, for twenty years chairman of the Brookwood Asylum Board, that efficient superintendence is rendered impossible in what he calls monstrous asylums, and that such institutions are most unfavorable to the treatment and cure of insane patients, and their management harassing and unsatisfactory to the medical superintendent. He says: "The doctors in large asylums cannot even know the patients by sight, much less by name. The thread of their history is to them a tangled skein which they scarcely attempt to unravel. The admixture of a curable patient with the vast common herd has a most detrimental effect."

Dr. Rayner of Hanwell complains of his large asylum. He says: "I cannot see all my patients in one day." And yet this very evil consequent upon overgrowth was at the beginning protested against by the English Lunacy Board. Its chairman, Lord Shaftesbury, said before the parliamentary committee of 1877: "My own opinion has always been, and from the first time I was acquainted with lunacy I have always maintained, that three hundred was the outside that could be well managed by one superintendent. There is a desirable process which the Germans call the individualizing system. With more than three hundred it would be impossible that the medical superintendent could personally see each patient, as he ought to do, several times in the course of the week. We had a very long, not to say very angry, controversy with Hanwell at the time that they enlarged their buildings. We were very averse to that scheme."

It seems to be necessary that every State should provide sufficient accommodation for its acute insane in small hospitals, where, under influences favoring restoration to health, recent cases could be made the subject of close study by skilful alienists, and where every possible means would be brought into requisition to effect cure, the ever-recurring question of expense being a secondary consideration. Only in this way can we expect to lessen the steadily increasing volume of hopeless insanity.

It must be conceded, however, that it has heretofore been found very difficult to establish small hospitals, or those even of moderate size, especially designed for and adapted to curative purposes. After such a project has been undertaken, and the Legislature has decided upon the site for a hospital, the work of building devolves upon a Board of which perhaps not a single member has made previous study of the special needs of the class to be provided for. An architect is employed who sets out to design an imposing building that will stand as an enduring monument to his architectural talent, rather than one best adapted for the care and cure of the insane. In his attempt to plan a grand edifice he is encouraged by those representing local interests, who desire a structure that will by its stateliness prove an object of admiration to strangers; and so, at the outset, we sometimes expend vast sums on immense buildings that have the semblance of palatial prisons. Later on, other difficulties arise. More accommodation is needed, and while it is plain that the interests of the insane require another asylum rather than additions to the existing one, pride in administering the affairs of a large institution, combined with influences arising from the benefits of local patronage, is powerful in swelling the accommodations, regardless of the purpose of the establishment and its original plan. Enlargement follows enlargement, until the institution finally reaches enormous dimensions, and whatever principles of cure were had in view at the outset, they are overwhelmed and rendered impracticable by the aggregation of numbers.

In the dominance of the architectural idea over the medical and moral in asylum building, we frequently behold a vast pile of brick and stone and mortar, in which are congregated a great number of mentally diseased persons, to every one of whom the surroundings are unnatural. Having brought them into abnormal conditions of living, the human mind is taxed to the utmost to overcome these and secure requisites to health, including a sufficiency of light and air. When we consider the distress of a patient placed in one of our immense hospitals built on the congregate plan, and subjected to the disquiet, unhappiness, and irritability consequent upon the herding together of masses of people, the wonder is that we should attempt his cure under circumstances that would seemingly cause a sane person to lose his reason.

The plan of a hospital for the acute insane I would have include a central, unpretentious building for the superintendent. This should also contain the offices and other apartments customarily belonging to what is known as the administration department. For the patients, there should be cottages so arranged that the superintendent's residence and central department will stand between the groups for men and those for women. The cottages should be of different sizes and variously designed, all of them resembling private dwellings. Assistant medical officers should reside in some of the cottages with the patients. There should also be accommodation in each cottage for the necessary attendants.

Reception cottages should be built in connection with every hospital for the acute insane. A fearful shock is felt by many patients when first brought into an immense prison-like building, and ushered into a long, formal hall or corridor furnished with angular wooden chairs and settees, ranged against the walls much like the waiting-room of a railroad station. Besides, the grating sound of strong locks, the sight of heavy doors, barred windows, and the strange-acting beings who gather around the new-comer are not likely to calm an excited mind. First impressions are lasting, and should be agreeable. They are more likely to be so, if the patient, when entering a hospital, approaches a pleasant cottage situated in park-like grounds, and, instead of the formal ward with its strange sights, is received into what is apparently an ordinary dwelling with homelike furnishings. Thus placed, the patient feels more at ease, and the physician can, for this reason, get at a better understanding of his case. Explanations as to what is for his best interest are made, his apprehensions allayed, confidential relations established between him and the physician, and, sooner or later, according to circumstances, he is transferred to the proper department. One thus entering a hospital for the insane, I am assured, is far more likely to be reconciled to his confinement, and, if curable, is sooner placed on the road to recovery.

In addition to reception cottages, there should be separate

buildings for convalescing patients. These should be like ordinary dwellings, and pleasantly situated. They should be comfortably furnished, and have every attraction that could agreeably engage the mind. It is asserted by a high authority that, when patients begin to mend, the process of cure is rapidly accelerated by removing them from the distressing associations connected with the first period of their disease to entirely new surroundings.

If food is conveyed from a general kitchen to the several separate establishments, as at Alt-Scherbitz, or if each cottage has its own kitchen, connecting corridors are not necessary. The latter plan presents the nearer approach to family life, and appears to be the more desirable arrangement for hospitals for the acute insane. In either case there is no arbitrary rule as to the relative position of the cottages and the central building, except that the former should not be too far away for convenient supervision. They may be located with reference to the grade of the land, to pleasant outlooks that may be had from them, or to picturesque effects. In any event, they should be placed far enough apart to give the appearance of retirement that may be attained by the judicious planting of trees and shrubbery between them. If corridors are considered necessary, they should extend backward from the central building and afford communication with such departments as are in the rear of it, and thence to right and left a short distance from the cottages, connecting with the rear of each of them by a branch section. An illustration of the way in which separate buildings may be thus connected is seen in the new asylum at Menstone, England.

Many of the old congregate asylums are so arranged that, to reach the most distant parts of the institution from the administration or central department, it is necessary to pass through every intervening sitting or day room on one of the asylum floors. Thus a monotonous tramping is kept up, which, on visiting days especially, is a constant source of disturbance to irritable and nervous patients. In every arrangement of buildings such an annoyance should be avoided. A common error in asylum building is made by connecting separate divisions, blocks, or dwellings by means of corridors at either end. It should be kept in mind that the best rooms are at the ends of a building, and they should not be spoiled by a corridor entrance at one end and exit at the other. If such passage-ways are deemed necessary, they should be so planned that persons may pass from the administration building to any one of the separate buildings without finding it necessary to enter any other than the one they intend to visit.

The problem how to dispose of the chronic insane so that they shall have suitable asylum care without interfering with the strictly curative functions of the hospital, has not yet been satisfactorily solved in England, Ireland, or Scotland; and the remedial powers of many otherwise good institutions, particularly in England and Ireland, are seriously interfered with by the excessive accumulation of chronic patients. The Visiting Magistrates of Haywards Heath Asylum, in one of their reports, say: " The Committee are strongly of the opinion that our county asylums are losing the character and the objects for which they were primarily intended, namely, as places for the sanitary treatment and cure of insane persons, and are becoming in a great degree mere receptacles for chronic and imbecile patients who are detained within their precincts; and this state of affairs greatly interferes with the due separation of the different forms of insanity so essential to the alleviation or the cure of brain disease." They further add that county or district asylums should be used "solely for the reception of acute and violent cases, of those requiring special treatment, and those affording the hope of cure ; and that provision should be made for the reception of aged and harmless insane persons in buildings specially adapted for their comfort and careful treatment, but constructed and conducted at a considerable less cost than our present asylums, and involving a much smaller expenditure to the rate-payers."

In Scotland, relief to a large extent has been found in the boarding-out system. In England and Ireland, a result of overcrowding in institutions originally designed for curative purposes has been to bring great numbers of the chronic insane under the unsatisfactory care of the workhouse.

The embarrassment arising from the overwhelming numbers of chronic insane in mixed asylums was very great in every country that I visited. A similar difficulty is becoming formidable in the United States, and there is no doubt but that it will continue to increase as the country grows older, unless a wise and comprehensive policy is adopted respecting this rapidly accumulating class. The weakening of the effectiveness of curative institutions by the continuous increase of chronic cases, and the insufficient provision for their care in pauper establishments, leads me to conclude that, in order to meet these difficulties, we must have more institutions of the character of asylums for the chronic insane, or for those who have received thorough treatment in a well-organized hospital. In the asylum, nevertheless, the hope of recovery should always be kept in mind, and the idea of incurability ignored. Such institutions should not be designated as chronic or incurable, but simply as *asylums* in contradistinction to *hospitals*.

As already stated, asylums should be located on considerable tracts of good arable land. The buildings should be simple and inexpensive. At the same time, they may be made attractive by varied outline and color. Adjacent to the administration department there should be sufficient hospital provision for those requiring special care. Beyond this, there should be a wide distribution of buildings on the colony plan. These, with their surroundings, should be made to approach as nearly as practicable to ordinary homes.

While there are conclusive reasons why hospitals for the acute insane should be small, these reasons have not the same force when applied to asylums for the chronic insane. In fact, there are some advantages that may be gained in bringing together a considerable number of this class. The opportunity is thus afforded of extended classification, of making up groups of artisans under a master workman, of purchasing supplies in large quantities at low rates, of reducing the cost of superintendence, and otherwise lessening the expense of maintenance. But the almost irresistible tendency to undue aggregation is a danger that should be carefully guarded against even in providing for the chronic insane.

The architecture of British asylums is without elaborate ornamentation. The prevailing opinion now seems to be in favor of two-story buildings for most classes of the insane. Even on the score of economy, little if any thing can be said in favor of a greater height, for although the same roof that would be required for a two-story building suffices for the three-story structure, the foundation walls of the latter are necessarily more massive than those of the former, the entire edifice must be stronger, and additional expenditures in these directions considerably increase the expense of a three-story building, so that its per-capita cost of accommodation is nearly or quite the same as that of a two-story structure. As an offset to the possibly slight advantage gained in the reduced per-capita cost of construction, there are the disadvantages arising from an extra flight of stairs, including the greater difficulty of escape in case of fire. The general form of modern construction admits of the upper portion being set apart almost exclusively for night use. The sleeping-rooms are unoccupied during the day, the windows open, and the bedding disposed for airing for a considerable time. By this arrangement the labor of the women can be utilized in putting in order the men's dormitories, and the change incident to separate day and night accommodation is conducive to healthfulness.

A mistake, and one which the builders of American asylums sometimes make, was occasionally found abroad in the failure to carry up the cellar or foundation walls sufficiently high to allow of a quick descending grade about the building, so as to carry off the water from storms that beat against the walls above. From this omission is likely to result damp foundations, which affect deleteriously the sanitary condition of living apartments.

The rule of the English Commissioners in Lunacy respecting air space in asylum buildings having ceilings twelve feet high is, that the associated dormitories for clean and healthy patients shall have fifty superficial feet of floor space to each bed, the separate sleeping-rooms at least sixty-three superficial feet, and that rooms occupied by sick or bedridden patients must have larger space with extra means of ventilation, that day-rooms must have not less than forty square feet of floor space to each person, and that a detached hospital for contagious cases must have fifteen hundred cubic feet of air-space to each bed.

The commissioners recommend that to each ward or

group of patients there be accommodation for two attendants, that their single rooms measure one hundred feet of floor space, and that, when practicable, their rooms, with glazed doors for observation, be placed between two associated dormitories; also that day-rooms for sick, aged, infirm, and excited patients should be on the first floor, and that no associated bedrooms contain less than three beds.

As affording greater security against fire, the commissioners advise that ceilings of rooms next below the roof be made of incombustible materials, and that buildings be semi-fireproof, or at least so constructed that a fire occurring in one part may be extinguished without destroying another part of the building; also that stairs be built of stone, with square landings, and the wall built up with hand-rails on each side. They insist that there must be a sufficient number of stairways, and that they be so placed as to afford ready egress in case of fire.

The commissioners advise that windows be large; that in day-rooms they be not more than three and a half feet from the ground, and in dormitories not more than four feet from the floor; that in single rooms they be stopped so as not to open more than five inches at top and bottom, and that a portion of them have strong, inside shutters so made that they cannot be forced open nor afford means of committing suicide by hanging; and that doors of single rooms open outward and fold close against the wall. They are of the opinion that for about one seventh of the patients in an asylum "infirmary" accommodation should be provided with abundant air-space, each room having an open fire or fires and a small diet kitchen; also that open fire-places should be built in all the large rooms, dormitories, and a portion of the single rooms, and that there should be other means of heating besides that of open fires. They recommend that there be a separate wash-house for very foul clothes, and a room for washing and drying the horse-hair of the mattresses; that there be an abundance of closet and store-room space, also conveniently arranged workshops well lighted and ventilated.

Some of the best English asylums are so constructed that the sewers do not enter nor pass under inhabited buildings, but terminate close to the wall in a ventilating flue after passing a syphon trap. In this way it is not possible for foul odors from the sewers to contaminate the atmosphere of living apartments. All soil and waste pipes are carried though the outer walls into the sewer and trapped. As an additional safeguard, the interior pipe system is also independently ventilated. The sewers are so constructed as to permit of convenient inspection at certain necessary points. It is thought objectionable to construct flues for ventilating drains or sewers in the walls of buildings. Lavatories, sinks, and water-closets are frequently placed in projections which are entered by a short corridor having good cross ventilation. Supply and waste pipes are of large size, and all those within the building are, as far as practicable, exposed to view. In some instances the waste from the baths and lavatories is discharged into automatic flushing tanks, and thus serves the purpose of flushing the sewers, from which surface water is excluded. Great care is taken to have sewers and drains laid on a solid bottom, equally graded from point to point.

For greater security against leakage and for permanency, as well as for better flushing in consequence of their superior strength, iron pipes are coming into use for sewers in place of glazed earthen-ware. Iron pipes are put together and secured at the joints with lead as are water-pipes, and when used, will answer to be smaller than those of glazed tile.

It is customary to have all pipes and sewers thoroughly tested before covering them in. Varnished cast-iron pipes are mostly used for water-mains, none of them being less than four inches in diameter. Wrought-iron service pipes are now commonly used in place of lead pipes because they are cheaper and more easily fitted. When laid under ground, however, they are incased in wood and protected by asphalt.

In many of the foreign asylums, in addition to bathing facilities for each ward or group of patients, there is provided for each sex a general bath-room, adjoining which is a pleasant dressing-room with open fire. Bath-tubs are conveniently placed at right-angles to the wall and away from it, so as to be accessible from all sides. They are sometimes placed in curtained stalls, or are completely secluded by curtains hung on rods. The floors and dados of bath-rooms, sculleries, and water-closets are generally of glazed bricks or tile, and on the floors of the bath-rooms is usually laid some protection for the feet. Turkish baths are not uncommon, and by some superintendents are thought quite beneficial to a certain class of excited and sleepless patients. In a few instances swimming baths similar to those devised by Dr. A. E. Macdonald of Ward's Island, New York City, are provided.

Inquiry was made at every place visited as to the comparative merits of plastered and unplastered interior walls. It was found in a number of large institutions that the walls were not plastered, but painted directly upon the brickwork. The opinion of experts, especially of those who had tried both ways, was largely in favor of the method of plastering. The unplastered walls are not so smooth, are less easily kept clean, and without great care is taken, form a harbor for vermin. Besides, to lay up interior walls in this way, better bricks are required, and more time is taken to lay them. If plastered on the brick, and afterwards painted or coated with silicate, the walls have a better appearance, and are easily washed. When a building is completed, little if any thing has been saved by not plastering.

It is pleasant to note the prevalence of cheerful open fires in the day and sitting rooms, the infirmary wards, and also in the dormitories of English, Scotch, and Irish asylums. In many ordinary wards these fires are without guards; in others they are partially protected, while in the refractory wards they are generally, but not invariably, screened by locked fire-guards. In the refractory wards at Hanwell, those in use were not locked. So far as I could learn, there had been an entire immunity from accident. Open fires are supplemented by some other method of increasing the temperature in severe weather, usually that of steam or hot-water heating.

Outer walls are frequently built with a hollow space between the bricks, as a protection against dampness. There are also flues with registers near the floor and ceiling, which may be opened or closed as occasion requires. These flues connect with horizontal air-ducts, communicating with a perpendicular shaft in which means of rarefaction are placed. In some places, as at Berlin, gas-jets are constantly burning within the wall flues, to facilitate ventilation.

The elaborate system of flues with great air-passages and enormous fans propelled by steam power, for forcing air into the various apartments, in use in some of our large asylums, was nowhere seen, either in Great Britain or on the Continent. In the construction of asylum buildings, we sometimes create at great expense a highly artificial condition which must be overcome by expensive contrivances. In no more important particular is this incongruity manifest than in the matter of ventilation. It would seem better not to depart so widely from natural ways of living, and build our asylums more as we build our homes.

Thermometers are kept in the various rooms and corridors of many British asylums, and the temperature is recorded at stated hours during the day and night. An incident showing the desirability of this regulation occurred just prior to my visit to one of the English asylums. The friends of one of the patients charged that much suffering had been caused by too low temperature in some of the rooms. The complaint found its way into the newspapers, and much feeling was aroused against the asylum. The management obtained the appointment of a committee composed of citizens and friends of the patient, to investigate the matter; and after a thorough examination, all were perfectly satisfied that the system disproved the charge.

In the construction of foreign asylums, the contingency of fire is not overlooked, and the various means of escape provided appeared to be sufficient. For protection against fire, British asylums have alarm signals, night patrols, the regularity of whose movements are recorded by tell-tale clocks, electric communication with alarm bells, hydrants conveniently placed on the different floors, abundance of hose ready for use, also a plentiful supply of buckets filled with water. The keys to the hydrants are close at hand, often locked within a glass-covered recess in the wall. By this arrangement the inspector can see whether the key is in its place; and it is easily accessible in case of emergency by breaking the glass. Some of the institutions have fire-engines with the usual appliances. A powerful steam force-pump in the engine-house connects with hydrants in every department, and in many asylums there is a fire brigade composed of employees, who are occasionally called out and trained. The attendants in each section are particularly instructed in precautionary measures against fire. In case of its occurrence, they are charged to look first to the saving of life and afterwards to the preservation of property.

FURNISHING AND DECORATION.

That nice appreciation of comfort which is a characteristic of the English people is noticeable in the furnishing of their best asylums. It is seen in the round-cornered furniture, the absence of angles and sharp corners in jambs, casings, etc., in cushioned and easy chairs for the infirm, the not infrequent use of carpets to deaden sound, and many conveniences elsewhere described.

It can be said of the asylums of Great Britain and Ireland that the beds are very comfortable. The bedding usually includes straw and hair mattresses, good pillows, and an ample supply of woollen blankets. For the refractory, the sheets are of very heavy linen strongly hemmed to prevent their being torn. It was observed that woven-wire mattresses were coming into use in the convalescent wards for quiet and orderly patients. It is important that the wants of the insane should be met for the night as well as for the day. Failure to provide them with good beds causes more or less sleeplessness and unrest during the hours allotted to repose, which results in disquiet and excitement the following day, in the worry of the attendants, and the disturbance of the orderly management of the whole asylum.

Not only is there a careful regard for the comfort of patients, but some degree of elegance is seen in the furnishing and fitting up of even pauper establishments in Great Britain. The wall decorations and the arrangement of artistic objects designed to attract attention and engage the mind are exceedingly praiseworthy. The interior embellishments at

Prestwich, Leavesden, Caterham, and some other institutions for pauper patients might, perhaps, be the subject of criticism were it not that the beneficiaries are insane persons for whom diversion of mind is necessary, and that these pleasing effects are produced by an ingenious and judicious utilization of the patients' labor.

Even old asylums are made to conform, as far as practicable, to modern ideas respecting an abundance of sunlight and cheerful interiors. There is no doubt but that many of the details of asylum construction and furnishing may be made to exert a healing and otherwise highly beneficial influence on the minds of the mentally diseased. Respecting the influence of bright interiors upon the insane, Dr. Clouston says:

"If persons are deprived of their reason and personal liberty, and taken away from their homes, too much can scarcely be done for their comfort and happiness, because nothing can possibly make up to them for what their disease has necessarily caused them to be deprived of. And there is in most cases of mental disease a tendency to degeneration in habits and ways, which it should be one of the most unremitting efforts of any good asylum, and all connected with it, to counteract. Nothing rubs off the veneer of good manners, tidy habits, cleanly ways, and all the little amenities and considerations for others that mark a civilized man, be he gentleman or not, so much as mental disease. Those who come in daily contact with the insane, seeing these things gone, would tend insensibly to treat them as if they never existed, or could not be restored, were not constant and strenuous efforts made to fight against the tendency. And if this feeling is given in to, it reacts on the attendant, and causes degeneration in him too. One means of counteracting this degeneration is undoubtedly by making the rooms and surroundings scrupulously clean and cheerful, the painting and wall papers bright and elegant, the carpets (if there are any) tasteful, and the clothing as good as the person would wear outside. If it is clearly seen that much thought and care are bestowed on these matters, down to the minutest detail, in any asylum for even the worst class of patients, it exercises an influence on them, and all who have to do with them, strongly counteractive of the lowering tendency I have spoken of. It greatly helps the moral and medical treatment. The whole spirit of the institution should be philanthropic and medical, not mercenary, prison-like, or merely disciplinarian."

GROUNDS.

The grounds of foreign asylums are, with few exceptions, elaborately laid out, and are under the superintendence of professional resident gardeners. In the arrangement of gravelled walks, a proper regard for the happy effects of broad green lawns is usually shown, and these were seen closely mown; while the walks and roadways were kept with great neatness. By means of propagating houses and by the removal of plants from the conservatories to the corridors, halls, and various apartments, and to the immediate surroundings of the asylums in summer, the institutions are most delightfully brightened and beautified. In planting, careful judgment is exercised, and trees and shrubbery are disposed so as to hide that which is unsightly, and to leave openings commanding the widest possible prospect of that which is pleasing. It was observed that walls inclosing yards were almost invariably hidden by shrubbery or creepers. In some places the haw-haw is used instead of a wall where it is undesirable to obstruct the view. The improvement of the grounds and keeping them in high condition afterwards is mostly the work of the patients, for whom it proves a healthful and agreeable occupation. The vegetable gardens are extensive, and, through careful tillage and the use of fertilizers from the asylum, are made very productive. The spade is largely used, and the farming is generally thorough and yields liberal returns.

In building an asylum much will be saved if, at the outset, a small piece of land is set apart for a nursery. By the time the grounds are so improved as to be ready for planting, the trees and shrubs in the nursery will be sufficiently matured for use, and thus afford a cheap and abundant supply of plants always at hand. The advantage of pursuing this course has been shown in the description of Brookwood asylum, where, under skilful management, the sterile heath was soon transformed into beautifully ornamented grounds.

ORDERLY ARRANGEMENT.

Great particularity is observed in nearly all the foreign asylums, especially in those of the British Isles, in respect to keeping every thing in its proper place. The storerooms attract attention by the orderly arrangement of their contents. Supplies are withdrawn from them only by requisition, and the books there are kept so as to show the stock of each kind of goods on hand. One cannot but admire the neatness of the linen closets, the tidy appearance of the dormitories, the clean aspect of the kitchen, the manner of putting away and housing tools and implements, and the care with which rubbish is assorted and stored.

It was frequently found that broken glass, scrap iron, lead, etc., had separate locked receptacles of deposit into which they could readily be put, but out of which they could not be taken except by use of a key. What is often regarded as waste material is saved and in one way or another turned to profitable account. In some places even the bones from the kitchen are pulverized, then dissolved and used for fertilizing the ground. Most of the asylums have a walled inclosure for depositing unsightly material only occasionally required for building and making repairs. By the use of these places of storage, every part of the grounds is made to present an orderly appearance.

SEWAGE.

The disposal of asylum waste was a subject of careful inquiry at every place visited. In the continental asylums, refuse was utilized for the fertilization of the soil by some of the various processes devised for this purpose. At several institutions liquid manure was conveyed to the land, while solid matter was converted into compost by being mixed with earth, chalk, and other substances. A plan highly spoken of was that of separating the solid from the liquid by means of a series of vats placed at different elevations in the fields, the waste being emptied into the highest, from which the liquid flows successively through the rest of the vats until entirely separated from the solid matter. What is termed the bucket system is common. The tubs or cylinders are daily withdrawn, emptied, and renovated. Their removal is effected from the outside of the building. In Great Britain some of the plans adopted on the Continent were in use, as also other methods more expensive. The English Commissioners in Lunacy recommend that sewage be distributed in a fresh state over the land, by gravitation through pipes, or by tubular drainage. Indoor portable earth-closets were used to a limited extent, and in some institutions had been rejected after trial.

Perhaps on no other subject investigated was there found a greater variety of conflicting opinions than that relating to the best means of disposing of asylum waste. Situation sometimes exercised a controlling influence, as in the case of an asylum located adjacent to a public sewerage system; but where this method of disposal was not available, widely different plans had been adopted, some of them quite expensive, and once it was found that one institution was introducing a plan just discarded by another. It would seem that where large agricultural operations are conducted in connection with asylums, the waste should be used to keep up the fertility of the soil and supply what is lost by cropping, as this is a natural process of compensation invited by the disinfecting properties of fresh earth.

GREATER FREEDOM.

There is a strong tendency in the better class of institutions to do away with many restrictions, which, at one time, were deemed absolutely necessary in the care and management of the insane. Especially is this noticeable in Scotland, in the abolishment of walled airing-courts, and the adoption of what is termed the open-door system. In some of the Scottish asylums one may walk through the establishment from one end to the other without unlocking a door. The windows on the first floor of what are termed "open-door" asylums, are not barred nor grated. They have sashes as in ordinary dwellings, and may be opened at will. It is said that the disuse of airing-courts in Scotland first came about by accident. An old asylum was undergoing repairs, and it became necessary to take down the walls of one of its airing-courts. The work of rebuilding being temporarily delayed, it was found meanwhile that it was possible to conduct the administration better without them than with them, and the cost of reconstruction was saved. It having been demonstrated that these prison-like barriers were both unnecessary and injurious, other asylums soon set about removing their enclosures.

The freer system, which prevails in Scotland, is extending in other countries. It is claimed by those advocating it, that the larger liberty given the patients does not add to the attempts to escape nor to the number of suicides, the tendency to self-destruction being thereby diminished. Dr. Joseph Petit, Superintendent of the Sligo District Asylum, Ireland, after having introduced the open-door system into the institution under his charge and having tried it for several years, speaks highly of its tranquillizing effect upon the patients, and regards it as a decided improvement. Respecting airing-courts, which he has abolished, he says:

"Airing-courts were provided to guard against the possibility of escape while the patients took exercise in the open air. In District Asylums these courts are usually so placed as to be bounded on two sides by the buildings. The atmosphere in them can hardly be called open air, and with regard to escapes, experience teaches, as might be expected, that the less the amount of restraint the fewer the escapes. Another very objectionable feature about airing-courts is the bad effect they have upon attendants, who fall into the error of supposing that when the patients are put within the walls of an airing-court they require no further looking after on the attendants' part. Judged from results, I think it will be admitted that money spent upon these enclosures is not only uselessly but injuriously expended. This last remark is also applicable to boundary walls, except where the grounds adjoin public roads."

So satisfactory has the absence of walled airing-courts

proved, that there is no probability of their being adjuncts to future asylums.

NON-RESTRAINT.

The discarding of old and cruel forms of restraint has been shown to be conducive to the recovery of the curable and to the comfort and happiness of the incurable insane. Freed from his bonds, with opportunities for recreation and employment, the patient, who, in former times, would have been a constant source of anxiety to those having him in charge, is now tractable, and even serviceable in lessening the pecuniary burden consequent upon his care. To maintain the system which produces this result, however, is a work that taxes all the ingenuity and resources of an intelligent and experienced medical staff supported by well-trained attendants. They must be ever vigilant to win the patient gradually to ways of gentleness if he be violent, and to arouse his energies and sympathies if he be melancholic.

The appliances for mechanical restraint were not found in foreign asylums to the extent expected. The crib was nowhere seen, and my inquiry for it, in some instances, was met by a look of surprise. Restraining chairs were sometimes observed, but muffs and gloves were only occasionally seen in use, and it was said that when they were put on, it was usually for surgical reasons. Padded rooms for the seclusion of maniacal patients were found in many of the British institutions. These are thickly cushioned on sections of plank, which are removable for purposes of renovation. The floor is not always padded, but is often covered with rugs of thick matting, as at Hanwell and Brookwood. The number of such rooms, always few, varies. In some parts they are losing favor. It was generally asserted that chemical had not taken the place of mechanical restraint. The terms non-restraint and seclusion, however, were very differently used, and I found at times that greater restrictions were resorted to than the correct meaning of these words would imply. As showing what some of the English asylum superintendents mean by restraint, attention is directed to the definition given by Dr. Brushfield, late of Brookwood asylum, in the notes on that institution.

From my observations in asylums in Great Britain and in this country, I should say that, on the whole, there was less restraint there than here, notwithstanding the fact that in many asylums in the United States it may be said to be virtually discarded. There can be no question but that the theory of non-restraint, once so thoroughly resisted, is now coming to be universally accepted; and the extent to which it has been adopted in recent years in our asylums leads to the belief that the time is not far distant when what is commonly understood as non-restraint and the open-door system will be put in practice to a greater extent in this country than it now is in Great Britain or on the Continent.

Some who approve of the system claim that under it the number of attendants is not actually or necessarily increased; others favoring it, concede that there must be a stronger corps of attendants, and that they must exercise greater watchfulness, but confidently assert that the benefits arising to the patient more than counterbalance the difficulties attending the carrying out of the freer system. It is claimed further that a quieter and more orderly administration is the result; that there is less destruction of property; that more patients are employed, and consequently there are larger industrial products.

In many of the British asylums it is the rule that, in case it becomes necessary to use personal force to remove a patient, a sufficient number of attendants shall be called in to accomplish the object without having a doubtful struggle. The fact thus made apparent, that opposition would be useless, frequently causes the patient to make no resistance whatever, and the desired change is effected without disturbance. An acute case, when violent and excited, is placed in exclusive charge of two experienced attendants, who give the patient several hours' daily exercise in the open air and watch him carefully in the wards until his excitement subsides and one person can assume the care of him. Finally, the other special attendant is relieved, and the case receives ordinary attention. The late Chairman of the English Lunacy Board in speaking of this method, said :

" It requires a greater number of keepers, or rather attendants, and it requires men of a very different character,—men of great forbearance and patience, men of great power of endurance ; for there is nothing on the face of the earth one half so provoking as a madman when he chooses to be so. I have looked in perfect astonishment at the character exhibited by the attendants in forbearance and moderation, and at what they patiently go through. To control a violent patient it requires three or four attendants. Formerly they would have put him in leg-locks and left him. The man would have become ten times worse, and the whole place would have been in disorder, because although you chained a man you could not stop his voice, and he roared and bellowed in such a way as few outside people can imagine."

MEDICAL OFFICERS.

The medical staff in foreign asylums is not so large as in similar institutions in the United States, and in some of them the number of physicians appeared to be inadequate. It was not found in any of the asylums visited that to female patients were accorded the medical services of physicians of their own sex, as is the growing and commendable practice in this country. The experiment, if such it may be called, of having two head officers to one institution, as was twice observed during my tour of inspection, has not worked satisfactorily in the opinion of those competent to judge, although no complaint was made by those in charge.

ATTENDANTS.

The success of asylum management is without doubt largely dependent on the maintaining of a good corps of intelligent and faithful attendants. One who has justly achieved distinction in the treatment of the insane says: "The longer I live, the more clear it is to me that good attendants, well trained, interested in their work, and proud of their success in it, with good heads on their shoulders, humane dispositions, pleasant manners, and ever using the brains they have to do their work, must be the sheet-anchor of success in an hospital for the treatment of mental disease."

An attendant who does not look upon a person mentally diseased with the same sympathy as he looks upon one bodily sick has a wrong conception of his relations to the patient, and is likely to be cruel when meaning only to be just. It should never be overlooked by those in charge of the insane that they are not responsible for their acts, and may be entirely unconscious of what they are doing. Failing to realize this, abusive language and personal indignities directed to the attendant awaken his resentment and a desire to discipline the patient. Hatred is thus inspired, and a permanent barrier is created between them. If the attendant would keep in mind the golden rule—Do unto others as ye would that they should do unto you—and imagine himself in the patient's place, and how he would like to be treated if similarly situated, much cruelty would be avoided and more of the insane would recover. The law of kindness is universal, and is as applicable in the treatment of the insane as in the treatment of any other of the helpless classes, and should be the guiding principle in their care. Though its influence may not be immediately perceptible, its subtle power gradually wins its way—producing quietness where there was violence and disturbance—and develops self-control in both attendant and patient.

As great suffering may result to the insane from neglect as from intentional cruelty or systematic severity. It is much easier to seclude or confine a man when restless or violent than it is to make some effort to employ him or divert his thoughts from real or imaginary troubles. Separated from the world as he is, it rests with the attendant to soothe and comfort; or, through indifference, incompetency, or acts of petty tyranny, to exasperate, and make the daily life of a patient unendurable, thus deepening the dark shadows that have gathered around his clouded reason.

Much has been accomplished in rescuing the insane from chains, gloomy cells, and scourgings; but the measure of reform in their behalf will not be complete until there is no possibility of their being subjected to the humors of ignorant, unfeeling, and incompetent attendants.

The need of persons systematically trained to properly discharge the duties of attendants is felt in Great Britain and on the Continent; but the effort to provide such by affording preliminary instruction does not appear to be so general as in the United States. In connection with a hospital for female patients at Morningside, there has been established a probationary ward and training-school for all the new-coming female attendants. Here they are taught to regard their patients as they would those of an ordinary hospital, and they enter upon their duties by learning to nurse the sick. Dr. Clouston says: "If any thing will produce a habit of kindness, this will be likely to do so." The plan of training at Morningside has proved of great advantage. In England, Dr. Wallis, of Whittingham, has instituted a system of training for attendants and nurses, and some other foreign asylums have entered upon the same undertaking.

In view of the large numbers to be employed and the peculiarities of the difficult service, the question of how the evil of incompetent attendants is to be reduced to the minimum is an important one. A partial solution will no doubt be found in the establishment of training-schools. By affording through these preliminary instruction, great advantage will be gained. Dr. Stephen Smith, the New-York State Commissioner in Lunacy, through whose earnest advocacy several training-schools for attendants have been established in the asylums under his supervision, says that where introduced there has been an improvement in the order and discipline; that patients are treated with more consideration; the sick are better cared for, self-reliance cultivated, emergencies more successfully met, the power of observation on the part of attendants quickened, and changes in both the mental and physical condition of patients more readily appreciated. The first training-school in the State of New York was organized by Dr. J. B. Andrews, Superintendent of the Buffalo State Asylum, and has proved eminently successful.

The object of the training-school being to prepare persons for competent attendants, those who have not the natural ability and cannot acquire fitness by education are of course weeded out. These schools should be so organized that certificates of graduation would be pretty sure evidence of the competency of those receiving them for the proper discharge of their duties. While the training should be both practical and theoretical, the instruction should not be so technical or difficult that the merely intellectual student would be advanced beyond those specially adapted by nature for their peculiar work.

It is gratifying to note that wherever opportunities have been afforded by the State in its training-schools for acquiring that knowledge so necessary in the care of the insane, a laudable desire has been shown on the part of attendants to increase their capacity for usefulness by the means thus provided.

The wages paid should be sufficient to induce those possessing more than average ability to enter the service. The plan adopted in England of yearly increasing the compensation of attendants till the maximum salary is reached stimulates them to faithful endeavor. The expectation that through serving long and satisfactorily, a government pension may be secured, is another incentive to a proper discharge of duty. The granting of pensions to superannuated officers and servants is permissible by the statute, in the county and borough asylums of England, the chartered asylums in Scotland, and the district asylums in Ireland.

Such persons should be employed as intend to make the care of the insane a permanent occupation. They should be selected from those who have attained mature judgment, but have not passed the prime of mental and physical vigor; they should have good health and cheerful and equable tempers; they should possess self-control, and be able to furnish proof of good moral character.

To further perfect and purify the service, the cause for leaving an asylum should be reported, in every case, to a central authority and made a matter of record, and no attendant should be employed without reference to the record. This practice obtains in England, where a register is kept by the Lunacy Commissioners of all discharged employees, and the cause of their discharge. In this country such a system could be made more valuable by co-operation among the different States.

The great strain to which faithful asylum attendants are subjected when on duty, makes it but just that they should have reasonable opportunities for relaxation and recreation when off duty, and that their accommodations and surroundings should be made pleasant. Their cheerful spirits, which are reflected upon those under their charge, should be fully sustained. Such arrangements would induce more of those best adapted for the service to enter it. This appears to be the view taken by the superintendent of one of the large asylums in England, where, as in this country, there is difficulty in securing and keeping qualified attendants. Bearing upon this subject, Dr. Ley of Prestwich says:

"The great problem in asylum management is, how to obtain good attendants; and when obtained, how to retain their services. In every asylum this difficulty, in a greater or less degree, has been felt, and in an institution of this magnitude, where obviously a greater proportion of experienced attendants is required, the difficulty in procuring and maintaining a staff of trustworthy subordinates has become a source of never-ending trouble and anxiety. No one conversant with the working of an asylum can doubt that much of the success of management, economical and otherwise, is dependent upon the character and reliability of the attendants, who are necessarily entrusted with the immediate care of the patients. The comfort, the safety, even the lives of those under their charge, depend upon the good conduct, fidelity, and watchfulness of these officials, who are, in point of fact, the instruments by which all the details of moral treatment are brought into practice. The service is an arduous one, and those who take to it are generally persons devoid of all training; consequently of the many who apply only a few are found gifted with the necessary qualities of temper and judgment, without which no good attendant can be made. . . . I think there can be no doubt that, apart from the question of salaries, much of the restlessness that affects the asylum attendant of the present day is due to the fact that his position is considered an inferior one, because the accommodation provided and the arrangements made for his comfort and relaxation are not equal to what persons in the same calling are able to obtain in other branches of the public service. In all the principal General Hospitals and Infirmaries it has been found necessary, in order to attract applicants of the requisite character and intelligence, to deal liberally with their nursing staff. Separate accommodation has been provided, and the comfort and convenience of the daily lives of these officials have been considered in every reasonable way. The result has been that the service is an attractive one, and hospitals and infirmaries have become serious competitors with asylums in the female labor market. I think it reasonable to expect that equal consideration for the comfort and accommodation of the attendants would be equally successful in rendering asylum service popular with candidates of character and ability, to whom the retention of their situation would be an object of some consequence."

NIGHT SERVICE.

The night supervision of some of the English asylums is very complete. The night staff is under a chief, and outdoor and indoor patrols are provided. Epileptic patients in their separate wards, as also the suicidal, are the objects of special and constant watchfulness. In some cases the attendants occupy a raised platform commanding a view of all those under their care. The beds for epileptics are ordinarily but a few inches from the floor. In the wards for the filthy particular attention is paid to the inmates, and the bedding is changed and the bath resorted to as occasion requires. In some of the foreign asylums, especially the French, a strict rule is enforced requiring every patient to observe a regular habit before retiring. Each attendant makes his report to the medical officer every morning and delivers his patients in a neat and satisfactory condition to the care of the day officer. The dormitories are warmed through the night by open fires, which aid in purifying the atmosphere. By means of electric communication assistance may be summoned in case of an emergency. The nightwatch is checked by tell-tale clocks.

RELIGIOUS EXERCISES.

A place for divine worship is found in connection with all public asylums in the British Isles. Although a hall is commonly provided for this purpose, sometimes there is a separate church edifice quite expensively built, with much attention given to architectural details. It is customary for men and women patients to enter from opposite sides and sit apart. A chaplain regularly officiates, and the attendance upon the service in some asylums is quite large. Arrangements are usually made for conducting both Protestant and Roman Catholic services, and at some places a Jewish Rabbi also officiates. It was generally thought that affording such of the insane as could properly attend divine worship an opportunity to do so, was only conceding to them a just privilege, and that the effect was salutary.

AMUSEMENTS.

The treatment of the insane in some of the leading asylums of Europe appears to be in accordance with the views of Dr. Conolly, who endeavored " to remove all causes of irritation and excitement from the irritable ; to soothe, encourage, and comfort the depressed ; to repress the violent by methods which leave no painful recollections, and in all cases to seize any opportunity of promoting the restoration of the healthy exercise of the understanding and of the affections." In the application of these principles, every variety of entertainment that can divert the mind or excite pleasurable emotions is brought into requisition.

A spacious, cheerfully decorated, and pleasant amusement hall, usually on the ground-floor, is considered indispensable in British asylums. A band, generally made up of asylum attendants, plays, at seasonable times, for the amusement of the patients. Besides concerts, dances, and other entertainments, there are cricket clubs, ball playing, etc. In some institutions an occasional costume ball is given, in the long preparation for which the minds of the patients are diverted. Their tastes in dress are gratified as far as possible. In these festivities benevolent persons residing near the asylum are encouraged to participate. Theatricals on an elaborate scale are also gotten up, from which those who attend derive much enjoyment.

Respecting dancing as a means of entertainment, the late Dr. Eames of the Cork District Lunatic Asylum said :

" Most essential of all amusements, I think, are the evening dances, which are held here regularly four nights weekly. All the patients look forward with pleasure when the day is over to these two hours of amusement in the recreation hall.

I was always of opinion that if this was beneficial to patients one night weekly, it was four times more so on four nights. On the evenings when there is no dancing, it is impossible to prevent the attendants allowing patients to retire to bed soon after supper; there is nothing for either to do, and time becomes irksome to both. When the patients go to bed at this early hour, they cannot sleep, and thus their night's rest is spoiled. On this account I have always insisted upon two hours being spent in the recreation hall on at least four nights in the week. Five hundred patients generally attend."

It may be questioned whether evening dancing and music are desirable so frequently as recommended by Dr. Eames, but numerous authorities favor some kind of entertainment on the evening of every secular day, in order to break the tedious monotony occurring between supper and bed-time.

At Morningside a special endeavor is made to afford the insane every possible advantage that can be derived from frequent and varied amusements and entertainments. Dr. Clouston says:

"The treatment of mental disease is in many cases a fight against morbid unsocial ways, degrading tendencies, and idle, selfish, listless, uninterested habits of mind; and we fight those by moral means, by employment, amusement, good food, fresh air, exercise, and good hygienic conditions of life.

"Amusements are, no doubt, of the greatest service in the treatment with a view to the recovery of insane patients. They are next to and supplementary to good food, cheerful quarters, kind treatment, medical care, and suitable employment. No one can be present at one of our dances without being impressed by seeing how a happy, cheerful state of mind takes the place of gloom in some patients through this amusement. They act on insane people somewhat in the way they do on sane people, but the effect is more marked from the sharp contrast between the insanity of expression and attitude before the dance and the sanity (so far as it goes and so long as it lasts) during the exercise. We have to provide a great variety of amusements to suit different tastes. No one should continue long enough or recur often enough to produce satiety. Different patients are amusable in very different ways and degrees too, just as different sane people are. There are many patients in such a morbid brain-condition, that they can scarcely be amused or interested in any thing, but those are the minority. Some of the patients get re-interested in the amusements to which they have been accustomed in their former lives; others are taken with new diversions. We train people to dance here who never danced before; to play billiards, bowls, cricket, tennis, curling, cards, dominoes, and draughts who had never played these games in their sane lives. I have known many cases where the interest in a game led directly to recovery.

"Most insane people, it must be admitted, do not rise to the enthusiasm which some games excite in ordinary people. The sanest *looking* people on our curling pond during an exciting game are *not* some of the officials. But that prince of all Scotch games has the power to rouse the dormant seeds of sane enthusiasm for the time in some of our patients in a wonderful way. An old curler seldom gets so insane that its enthusiasm and its familiar terms don't come back to him when he finds himself on the ice. This year (1885) we had a very exciting struggle for the silver medal, which Dr. Batty Tuke, an old assistant physician here, most thoughtfully presented to our club. It was felt by all that day that the best curler was the best man, quite irrespective of his social or mental condition. The day was fine, the ice good, the pond was surrounded with crowds of interested onlookers, and some of the ladies stimulated the players by the very outspoken expressions of their hopes as to whom they would like to see successful. One man, a keen and fine old curler, had been for some time in a dull humor. He curled in a dogged but apparently uninterested, unenthusiastic way, not speaking to any one. Towards the finish his score ran up fast, and he did begin to look a little more lively; but when with his last stone he 'chappit' the winner, and so beat the man by one point, who had been running him neck and neck, the old curler's spirit triumphed over the insanity, a sudden change came over him, and he became 'another man,' throwing his 'besom' into the air, curler's fashion. His whole expression and attitude changed, as he received the congratulations of his friends with smiles, and made an eloquent and appropriate speech on getting the medal. I could not help thinking that if I had some medicine in the surgery which would take hold of my patients' brains, as curling had done in this case, our recovery-rate would be a larger one."

EMPLOYMENT.

Many foreign alienists lay much stress on the efficacy of wisely directed labor in producing a tranquillizing influence upon cases that seem amenable to no other kind of treatment. Notable illustrations of the beneficial effects of laundry and farm work have been set forth in the descriptions of Alt-Scherbitz, Clermont, and Woodilee. At the last-named institution from seventy-five to eighty per cent of the inmates are said to be employed usefully and profitably. Here the presence of skilled artisans and experienced farmers and gardeners, not only directing, but working with the insane, has been productive of excellent results.

The most acceptable and healthful occupations being those of gardening and farming, the English and Scotch have an advantage over us, in a climate where it is possible to send their patients out-of-doors to work or take recreation during the greater part of the year. In Scotland the benefits of agricultural employment appeared to be more highly appreciated than in England, nearly all the men capable of so doing working out-of-doors; while in England they were frequently seen in pleasant weather engaged at indoor work, such as ward cleaning, bed making, etc. The time daily devoted to outdoor labor at Woodilee was from one and a half to two hours longer than in some of the asylums near London. It was the common opinion that abundance of outdoor recreation and employment was the secret of tranquil nights and contentment, which lead to recovery.

A variety of employment should be provided, so that patients can be placed at work congenial to them. Dr. Williams, of Haywards Heath, advocates in certain cases a change from the occupation to which the patient has been accustomed to one with which he is unfamiliar. He says: "A townbred artisan, such as a tailor, admitted in a low state of bodily health, if sent out daily for a little gentle employment in the garden, generally becomes greatly improved physically, and concurrent with the physical improvement often comes gradual restoration to reason. So also a rustic, tending towards dementia, if put to work at a trade, often brightens up considerably; whereas, if sent out to work on the farm or in the garden, employment that he had been used to from birth, would set to work at it and perform it mechanically, and it would be of but little benefit to him."

The laundries, in which large numbers of the insane in foreign asylums are employed, are conveniently arranged, and have goodly sized rooms for assorting, mending, and folding clothes. Although there is usually sufficient machinery for washing, there seems to be a growing tendency to do more of the work by hand, in order to make available a means by which excitable patients may work off their superfluous nervous energies. For a certain class of female patients hand washing has been found to be the best of all employments, and may be used to a large extent as a substitute for seclusion and restraint. In one asylum an entire plant of washing machinery had been set aside and hand washing substituted for remedial reasons. It may be well to consider whether this course might not be pursued with advantage in other asylums where machinery is used. The experience of those institutions in which the washing is done by hand seems to warrant such a course. In the abstract, washing by machinery is cheaper than by hand, but when the number of attendants that may be dispensed with in the refractory wards of an asylum having a large laundry operated by hand is considered, it is claimed by some that the last-named method is cheaper.

The English Commissioners in Lunacy lay great stress upon the desirability of providing occupation for the insane. They say: "In the treatment of the insane great importance should be attached to the subject of their useful employment. Our aim constantly is to encourage the efforts of superintendents to devise suitable occupations, and to induce their patients to engage in them; and with the view of ascertaining as nearly as we can the extent to which such efforts have been successful, we have instituted a comparison of the results attained in the years 1877 and 1886 respectively." The Commissioners found that fifty-seven per cent of the patients in the county and borough asylums were employed in 1877 and sixty-two per cent in 1886. In some English asylums the proportion employed during the latter year was as low as forty-five per cent and in others as high as seventy-eight per cent.

The custom of inducing patients to engage in useful employment is extending to hospitals and licensed houses, and the efforts made in this direction are encouraging. At St. Andrew's Hospital in Northampton, out of 160 gentlemen at one time in the institution, 56 were occupied in the summer at farming and gardening, while 18 worked at carpentry, printing, etc. Respecting the beneficial effects of outdoor employment, Dr. Lloyd Francis, who is connected with this institution, says:

"Outdoor labor is looked upon as a therapeutic means of the highest possible value, and each year adds fresh evidence of its efficiency. It is put to trial in one stage or another of every case where physical disease or extreme exhaustion do not contra-indicate. The means of persuasion are necessarily more limited than in a county asylum; the bait of certain privileges and small extra luxuries—ale, tobacco, and the like,—so tempting to the pauper, cannot lure the private patient, whose diet is ample and varied, and whose material comforts no amount of industry can increase. There remains, then, only argument, moral suasion; and hence oftentimes much difficulty in overcoming the irrational scruples, more especially of young men. The idea of digging, road-making, or wheeling a barrow is, even in the guise of medical treatment, at the outset, rather shocking to the school-boy, the undergraduate, the lawyer, or parson. He resents the proposal as an outrage to his dignity; declares that he was sent here for rest and remedies, not to do laborer's work, that such toil is all very well for poor people, but not for gentlemen—that, in short, he will have none of it. His repugnance, however, generally yields in time to reiterated advice and the example of others, and, once vanquished, seldom revives.

"The result, in the immense majority of cases—I might

say in all,—is beneficial. Over and over again do we note instances of rapid and complete recovery following steady application to outdoor work, when other means have signally failed and the prognosis has become decidedly bad, and, coincidently with the mental improvement, the establishment of physical robustness and vigor such as the patient has not often previously possessed. Such patients have the fresh, ruddy complexion, fat cheeks, and happy contented aspect which one observes in convalescents from typhoid fever."

As affecting chronic cases and respecting the comparative merits of employment and amusement, Dr. Francis says:

"In the treatment of chronic insanity, too, outdoor employment, though of necessity rarely curative, is yet of unquestionable value. A chronic lunatic of the worst type —turbulent, noisy, destructive, treacherous, violent, faulty in habits, an inveterate nuisance—shows marked improvement after a few months of steady work. Sleeplessness which drugs have failed to influence yields to healthy fatigue; he no longer makes a scarecrow of himself by tearing his clothes; his opportunities for self-abuse are much curtailed; and he relieves his angry feelings by vicious digs into the earth, or kicks at his barrow in place of murderous attacks upon fellow patients or attendants. Finally, his appetite is more keen, his food better assimilated, and his general health improves. Now and then such a patient even attains a state of fairly permanent partial recovery—a placid, contented, feeble-minded condition, it is true, but still enviable in comparison with his former miserable existence.

"It may be asked whether outdoor amusements with athletic exercises would not more agreeably serve the same end. The answer must be in the negative. For one patient who is capable of taking part in outdoor games, at least twenty can be put to manual work. An acute maniac or a dement can be made nothing of on the cricket field or tennis lawn, though he may dig or break stones with energy and purpose. Moreover, field sports and athletics are apt to be indulged in spasmodically—a few hours of violent exercise and excitement, followed by a long interval of rest and indolence. . . . The groundwork of recovery is found by experience to be best laid in steady plodding within the hospital boundaries; later, when convalescence is fairly established, play may safely vary the monotony of work, or even be substituted for it, though frequently such a patient, recognizing, as he improves, what a good friend work has been to him, goes on quite contentedly."

For medical reasons, if for no other, it would seem that a rule requiring boarders or private patients, as well as paupers, to be in some way occupied or employed, might be made applicable in every case where the condition of the patient permits. The ennui and disgust arising from an idle, monotonous daily routine of life, even with luxurious living, are almost intolerable to a sane person, and in the case of the insane, who are restricted to narrow limits, frequently lead to unhappy results. The observance of the foregoing rule has proved so beneficial, where enforced, that it should be regarded as one of the leading principles of treatment in all hospitals and asylums for the insane.

In providing employment, the welfare of the patient, and not the pecuniary advantage to be derived from his labor, should be kept constantly in view. It may be necessary to exercise as much watchfulness to prevent some patients from over-exerting themselves as tact to induce others to engage in any occupation. The hours for work should be limited, the whole industrial system carefully supervised, and the instructors or attendants should not act as overseers, but as companions, working with the insane, thus dignifying labor by their example. In every case due regard should be had to the mental and physical condition of the patient, which should be ascertained, as is done in some asylums, by a medical inspection made every morning.

REMUNERATION.

The custom of making a small remuneration or granting special favors or privileges for work performed by the insane, as practised at Fitz-James, Gheel, and some other places, is spoken of with much satisfaction by superintendents who have tried the plan. It is based on the belief that, although a man's mental faculties are impaired, he may still have understanding enough to appreciate an act of kindness and justice. On some points he may be perfectly rational and intelligent, and may know and sensitively feel, as well as any of his sane brethren, that labor should have its reward.

In the Government Asylum for insane criminals at Broadmoor, England, the plan of making money payments was introduced in 1877, and after a fair trial, has proved highly beneficial to the inmates and advantageous to the asylum. The value of the labor performed is estimated by the hour or by the piece, one eighth being credited to the patient, which he may expend in harmless luxuries or have sent to his family. Each has a book of forms in which his requisitions are entered. The sum paid to the workers amounts to about £300 yearly.

In France, the principle of making moderate money rewards to patients who work, is sanctioned by the Government. The Inspector-General of Insane Asylums in that country is of the opinion that patients should receive some slight recompense for their labor, and that this might be in the gratification of some pardonable whim or in a money payment on leaving the asylum if cured. He says: "We have here a moral means which acts powerfully on certain natures. With others the effect is not so marked, but were no other result obtained save that of provoking a childish joy, it ought not to be despised." The French system is thus officially described :

"Inasmuch as work is considered a therapeutic agency, it is necessary to encourage patients to devote themselves to it with a certain amount of assiduity. All work deserves pay. The greater part of the insane are aware of that, and insist that their work should be rewarded. From the first month a certain proportion is allotted by the physician in charge to a fund for the patient on his discharge. When the discharge fund has been attained, the surplus goes to the profit of the patient, to be devoted to Sunday dress, to articles of fancy he may desire to purchase, or to remit in whole or in part to his relations. All these favors become in the hands of the doctor a therapeutic agency of a certain importance, and they are a satisfaction to the patients, which it is right to afford them."

Compensation would do away with the sense of injustice felt by many of the insane in being obliged to work while confined against their will. Numerous complaints, in some cases very bitter ones, have been made to me by patients in insane asylums because they were not paid any thing for their services, and had no privileges over non-workers and idlers. Some recompense would help to make such persons contented, and their cheerful spirits would animate others. Dr. Mitchell, of the South Yorkshire asylum at Wadsley, says that much more might be done for the improvement of the mental and bodily health of the insane if a greater number could be induced to engage in labor. In his experience, the lack of compensation to workers stands in the way of a complete industrial system, many persistently refusing to work, on the plea that they have never worked without remuneration.

DRESS AND CLOTHING.

Foreign asylums are usually kept at a low temperature, and the degree of heat maintained would, in this country, be considered insufficient for the health and comfort of a large class of patients. The opinion, however, was generally expressed by the superintendents of asylums in Great Britain and Ireland, that it is far better to plentifully provide warm and substantial clothing for the patients than to keep their rooms at a high temperature. In the latter case they claim that the inmates become more sensitive to the cold when exposed, and consequently cannot derive the great benefit which may be secured by regular outdoor exercise. Although the winter climate of England is less rigorous than ours, the clothing of asylum inmates seemed to be more substantial and comfortable than that furnished patients here. Stout shoes and warm woollen socks or stockings are worn. High-collared, heavy cloaks or capes, buttoning in front and extending over the shoulders and arms are provided in some of the asylums for patients taking recreation or country walks in raw weather.

Many asylum superintendents deprecate uniformity in dress, and it was pleasing to observe that in the selection of articles worn by patients, such as shawls, capes, ribbons, bonnets, etc., their preferences as to color and pattern were frequently gratified.

DIETARY AND DINING-ROOMS.

In some asylums in England and Scotland, where male and female patients dine simultaneously in the same hall, the practice is approved by superintendents, on the ground that it is more in accordance with the ordinary way of living. At the small county asylum at Haddington, it has long been the custom for men and women to dine even at the same table, and no inconvenience has resulted therefrom. Attendants usually preside at the ends of the tables. As a general rule, it is desirable that tables, each accommodating a limited number of classified patients, should be provided in the dining-halls.

Knives, forks, and spoons are used in the more modern institutions. The knives for the refractory have frequently blunt edges, with the exception of about an inch and a half near the point. In Sweden, wooden spoons are provided for this class of patients. In most of the large asylums visited, the food is conveyed to the dining-rooms from a general kitchen.

It did not appear that the variety and quality of food common to American asylums was equalled abroad. In the institutions for the dependent insane, the difference was quite marked, the foreign dietary being far less generous. The food is possibly as good as the daily fare of agricultural laborers in the districts from which patients come, but mechanics from the cities find it inferior to that to which they have been accustomed. In the Scotch institutions the acceptability and general use of the various preparations of oatmeal promote the healthfulness of the patients and at the same time tend to reduce the cost of maintenance. Much less fresh beef is consumed in foreign asylums than in those of the United States, and there is not the variety of vegetables nor the abundance of palatable fruits. It should not be forgotten that the insane need a nourishing and liberal diet. If, with this, they are well employed, they improve in health and become more orderly and quiet.

ALCOHOLIC STIMULANTS.

An American regards with surprise the extent to which alcoholic stimulants are used in foreign asylums. In Great Britain and Ireland and some other parts of Europe, their use is gradually but surely decreasing. It is computed that the quantity consumed in county asylums in England diminished fifty per cent in the seven years preceding 1883, and nearly seventy-five per cent in borough asylums during the same period. One of the managers of the Brookwood asylum asserts that where the use of alcoholic stimulants is lessened, fewer drugs and narcotics are found necessary; and that the substitution of milk in large quantities for beer, ale, and spirits is attended with the best results. The English Commissioners in Lunacy state that in English county and borough asylums the amount of surgery and dispensary expenditures is lowest where the consumption of wines, spirits, and porter is smallest. A superintendent of one of the English asylums assured me that the giving up of beer in his own asylum, except for medicinal purposes, had proved beneficial, the patients usually eating more, and, in a majority of cases, gaining in weight. In the summer of 1884, Dr. D. Hack Tuke, in gathering statistics on this subject, found that, out of one hundred returns made to him from different county and borough asylums and registered hospitals, one half showed the non-use of alcohol except as administered medicinally, the officers generally reporting that the discipline of the asylum had at the same time improved.

SOCIAL DISTINCTIONS.

The attempt to make a classification based on social standing is a cause of great embarrassment in some of the foreign asylums. The inconvenience resulting therefrom, was strikingly apparent at Hamburg. At Halle, even in the small cottage hospital for sick women, provision was made for two classes. The arrangements necessary to carry out such a system, increase the cost of care and lessen the available capacity of the institution, as there may be lack of space for one class while there is an excess for another. The observance of these social distinctions also stands in the way of a proper classification based on the mental and bodily condition of the patients.

VISITATION.

In Great Britain and on the Continent the evil of excessive visitation by habitual sight-seers and by people whose motives are simply to gratify an idle curiosity, is not so great as in some parts of the United States. While asylum visitation is free to those in authority, the admission of the public is subject to rule. Even pauper institutions enjoy comparative seclusion. At Haywards Heath, and also at some other asylums, friends are not allowed to visit a patient until a month after the date of his admission, and they are then admonished to say nothing that may depress or disturb him.

It would seem that to the misfortune of insanity should not be added the mortification felt by a patient in consequence of public exposure while under care and treatment. It sometimes happens in our asylums that the strange delusions of an insane person are treated with idle jest or thoughtless ridicule by visitors, and his excitement thereby increased. Visiting by relatives and those actually interested in the insane should be permitted under proper statutory regulations, to such an extent as will preserve the confidence of the public in the asylum and further good administration; but to exhibit patients to curious visitors and point out their grotesque characteristics, as is sometimes done, in much the same way that animals are shown in a menagerie, cannot be considered other than an outrage to them and an injustice to their friends and relatives. It should not be forgotten that, in an asylum, the patient is entitled to as much of his former privacy as is compatible with his peculiar mental condition.

LETTERS.

Respecting the disposal of letters, always a source of more or less embarrassment in asylum management, an act of the British Parliament, passed in 1862, requires that in England every letter addressed by a private patient to the Lunacy Commissioners must be forwarded to them unopened. Those addressed to others are left to the discretion of the asylum superintendent. Such as he thinks should not be mailed must be brought to the attention of the Commissioners in Lunacy, the Visitors, or the Committee when making their next inspection. The opinion of the late Chairman of the English Lunacy Board respecting the detention of letters was given to the special parliamentary committee of 1877 in the following language:

"I am inclined to think that the correspondence is very fairly carried on; that letters that ought to be sent are really sent by the great mass of superintendents. I cannot have a doubt that it is so. All letters that are not sent are reserved for the Visitors or the Commissioners at their next visit. On inquiring the other day I found there were very few instances indeed in which the Commissioners or the Visitors upon opening those letters thought they ought not to have been kept back. No doubt the superintendents of asylums have very great power, and they might keep back a great number of the letters that ought to be forwarded.

I do not think they do it, and I do not think they are inclined to do it: the responsibility is very serious. I do not think they keep back any correspondence but that which they think would be positively hurtful; and such must, of course, be detained. Some of it is of the most blasphemous and obscene character. . . .

"There are some patients who have a positive frenzy for writing. I have a correspondent in Sussex House who favors me periodically with some of the longest letters I ever saw. They are invariably sent to me, and I do myself the honor—I cannot say the pleasure—of reading them. The man boasted to me the other day that he had written no less than one hundred and twenty letters within the last month. You can hardly think that all those could be forwarded."

The statute regulating the transmission of letters in Scotland is as follows: "Every letter written by a patient in any asylum or house, and addressed to the Board or their Secretary, or the Commissioners in Lunacy, or any of them, shall, unless special instructions to the contrary have been given by such Commissioners, or any of them, be forwarded to its address unopened; and every letter from the Board or their Secretary, or such Commissioner or Commissioners, to any such patient, when marked 'Private' on the cover, shall be delivered to him unopened; and every person who shall intercept or detain or shall open any such letter without the authority of the patient by whom it is written or to whom it is addressed, shall be liable in a penalty not exceeding ten pounds: provided that the Board shall transmit a copy of such letter to the superintendent of such asylum or house if it shall appear to the Board that the contents of the letter are of such a nature that it is of importance that the superintendent should be made acquainted therewith." The superintendent exercises his discretion in regard to other letters.

In accordance with the law of Massachusetts, locked boxes are placed so as to be accessible to the patients in asylums; and into these, letters can be dropped, which must be forwarded unopened to the Commissioners of Lunacy and Charity. The law has been in operation since 1874, and the opinion of one intimate with its workings during that time is valuable. Mr. F. B. Sanborn, Inspector of Charities for Massachusetts, expresses himself in favor of the law. Although the great mass of letters that find their way into the boxes, he says, are little else than rubbish, by their use a legalized means of communication with the outside world is secured, which affords satisfaction to many of the insane and possibly to their friends.

It would seem that the law should provide some means whereby State supervising authorities and patients can communicate directly with each other, as in Scotland, and that all letters not forwarded to friends of patients by the asylum superintendent should be held for the inspection of the supervising commissioners, whose duty it should be to examine the same when next visiting the institution.

POST-MORTEM EXAMINATIONS.

There is no uniform rule as to the making of *post-mortem* examinations in the English county and borough asylums. The general practice is that of communicating with relatives or friends in the event of the death of a patient, sometimes stating that an examination will be necessary; and if there is no objection, an autopsy is made. At St. Luke's Hospital it is set forth in the printed form of admission that, unless objection shall have been made in writing to the asylum secretary previous to the decease of a patient, a *post-mortem* examination will be held. In some institutions very few are made ; while in others they are made in nearly every case. During the year ending January 1, 1887, out of 5,053 deaths in the English county and borough asylums, 3,649 *post-mortem* examinations were made. The practice is thought desirable by the English Commissioners in Lunacy whenever possible, and with the observance of the same rule as to the consent of friends or relatives of all classes of patients. Besides the advantage to be derived from these examinations in advancing the knowledge of the pathology and treatment of various forms of insanity, the commissioners deem them desirable as a means of discovering injuries, and think the evidence of possible maltreatment that may be thus obtained is a protection to the insane.

ADMISSION AND DISCHARGE.

While every possible precaution should be taken to guard against violations of personal liberty, the process of entering an asylum should not be so hedged in with needless forms as to prevent the speedy admission of the patient. The highest medical authorities concur in the opinion that early treatment of insanity is essential to the attainment of successful results, and that loss of time is disastrous by making permanent the departure from normal action which, under favorable conditions, might have been quickly restored. To facilitate the early entrance of an insane person into an asylum, no more publicity should attend the transfer than is necessary. An objection to the system of trial by jury is that the dread of public exposure incident to this mode of procedure frequently delays action by friends of the patient until special remedial measures come too late. Besides, the excitement experienced by the patient when forced to appear before a court and jury as in a criminal process, confronted by adverse witnesses whom he had hitherto regarded as trusted friends, has often a very damaging effect upon him.

The dividing line between sanity and insanity is usually so imperceptible that it is difficult to fix the precise period when a person enters the first stage of mental disease. Dr. James Coxe has truly said: "As day passes gradually into night, so does sanity pass gradually into insanity." In beginning treatment we should therefore come as near to this uncertain line as practicable. It is better that we should err by sending the sufferer to the hospital too early rather than too late, always providing for easy means of egress as soon as he is in a condition to justify his leaving. When the nature of insanity and the necessity of special hospital treatment are generally well understood, when the laws governing admission and discharge come to be perfected, and when other desirable reforms are effected, there will be little if any more reluctance on the part of friends to place a patient in a hospital for the treatment of mental than one for the treatment of bodily disease.

It is important that the relations between the patients and the superintendent of an asylum should be of the most friendly nature, and that the latter should enjoy the unreserved confidence of his wards. This cannot be the case if the patient thinks that he is entitled to his discharge, but fancies that it is withheld through the selfishness or caprice of the superintendent. Feeling that a tyrannical power is exercised over him, the patient is irritated whenever he sees his imagined oppressor. It has been my frequent experience, in passing through the wards of an asylum with the medical superintendent, to hear, on every side, entreaties to and maledictions against him, the inmates believing that he held the key to their freedom and wrongfully detained them. It would appear that the power to grant discharge should be vested in other authority than the superintendent. Should he be relieved of this responsibility, he would be placed on a more friendly footing with his patients and be able to administer more effectually to their recovery.

As a safeguard against unnecessarily prolonged detention, it would seem desirable that, after a patient had been in an institution three years, an affidavit as to the necessity of his further care therein should be made yearly by the authority in which is vested the power of discharge. This, although perhaps of little account as affecting the numbers released, would doubtless be a satisfaction to friends of patients and to the public.

The power exercised by the Scotch Lunacy Board to discharge upon a certificate of recovery granted by two physicians whom they have called to make a special examination, is a very satisfactory way of meeting the question of discharge in controverted or doubtful cases. The power vested in the sheriff (county judge) to discharge upon a proper certificate presented to him by two reliable medical men, to the effect that the patient is either recovered or is not dangerous to himself or the public, makes the exit from an asylum in Scotland easy and encourages resort to early asylum treatment.

The granting of probationary discharges,—that is to say, permitting a certain class of patients to go out on trial to their homes, to work for wages, or, under the guardianship of friends, to the mountains or the sea-side, appears to be attended with good results. The practice is in favor with the Scotch Lunacy Board, and in England it is becoming more prevalent.

In the disagreeable task of taking an insane person to an asylum, the relatives or friends should only participate when their doing so would be helpful to the patient; and they should endeavor to show him that his going is unavoidable, and that they deplore its necessity. Strategy and deceit are frequently resorted to in making these transfers, and it is an embarrassing task for the superintendent of an asylum to undeceive the person brought to his care under such circumstances. He is thus placed in an equivocal position, and at the outset is looked upon with distrust by the patient, who should, for his own sake, have entire confidence in the one with whom he is brought into such intimate relations and who is to control his future destiny. An insane person should never be deceived, but first, last, and always told the truth. The burden of his complaint is, that he is the victim of a conspiracy among his friends. Believing himself to be sane, how could an insane man feel otherwise than deeply wronged at the treachery of those whom he had always trusted and loved? His worst passions are aroused, and a sense of desolation and bitterness adds to the tumult of his disordered mind. In the transfer of female patients to asylums, the statute should require that they have the protection that the companionship of a competent female attendant affords. For obvious reasons, all transfers should be so arranged as to avoid, if possible, reaching the asylum at night.

VOLUNTARY PATIENTS.

In England, a person that has once been an inmate of an insane asylum can, by payment of board, be re-admitted at his own request, if this shall be made within five years after his discharge. The procedure respecting the admission of voluntary patients into asylums in Scotland has already been described. So far as I could learn, the operation of the Scotch law relating to this class of patients had proved satisfactory. In 1881, Massachusetts passed a law permitting persons to enter asylums voluntarily. During the five years after the enactment of the statute, there were 261 admissions under it, thirty-five of which were re-admissions. All of those admitted except forty-two had, on January 1, 1887, been discharged. During the year 1886, about one third of the admissions to the McLean Asylum, Massachusetts, were voluntary. The admission of this class of patients into insane asylums tends to do away with the air of mystery and feeling of dread with which these institutions are sometimes invested in the public mind.

SUMMER RESORTS.

It has been shown that some of the foreign asylums, notably Morningside and the York Retreat, take their patients in the convalescing stage, at seasonable times, to sea-side resorts. This praiseworthy custom is also observed on this side of the water by the McLean Asylum in Massachusetts and the Friends' Retreat in Pennsylvania. The advantages sought may also be secured at other places than the seaside. Attractive points upon the shores of our beautiful lakes and near our magnificent rivers, or among the mountains, would afford the desired change. Judging from the beneficial effects which have resulted from a change of scene and air, it would seem advisable to extend the privilege as far as possible to all patients likely to be benefited thereby. The removal need not involve a large outlay, as suitable premises could be secured by rental, and temporary structures, or even tents, could be made to answer the purpose of shelter. Such resorts should not be called by any name suggestive of insanity.

CRIMINAL INSANE.

Separate provision should be made by every State for insane criminals. The necessity for this is so apparent, if considered only with reference to the rights of the non-criminal insane, as seemingly to need no argument. What has been done by the English Government at Broadmoor and what New York State is now doing under the intelligent supervision of Dr. Carlos F. MacDonald, Superintendent of the State Asylum for Insane Criminals, demonstrate the advantages of providing separate establishments for the criminal insane.

BOARDING-OUT.

The number of insane provided for in private dwellings in continental countries aside from Belgium is not easily obtainable. It is doubtless considerable. There is a large number in family care in Ireland. On the 1st of January, 1887, there were in England 5,809 pauper lunatics "not confined in any asylum, hospital, or licensed house, but residing with relatives or others as outdoor paupers." These formed 8.02 per cent of the whole number of pauper lunatics. As idiots are included in the term lunatic, what proportion were insane cannot be determined from the statistics given. During the ten years preceding 1887, the number of outdoor pauper lunatics in England decreased 503. There is no general system for boarding out the insane under governmental supervision either in England or Ireland.

On the 1st of January, 1877, the insane boarded in private dwellings in Scotland numbered 1,418 and a decade later there were 2,140, being an increase of 722. This increase may be partly due to the protection afforded by the watchful supervision of the Scotch Lunacy Board—a kind of supervision which is essential to the success of the boarding-out system. The opinions of this Board already given and of the officers connected with it, who bestow special attention on the important work of visiting the insane placed under family care are worthy of careful consideration.

It is not improbable that, if a searching inquiry were made into the state of the insane in private dwellings in the several countries except Scotland, an unsatisfactory condition, showing much abuse and neglect, would be revealed.

Among the chronic insane in American asylums there are doubtless many harmless lunatics who could be removed from asylum to family keeping, if a proper and efficient system of supervision were extended over them. As the country grows older and the dying out of industries in certain localities causes the people to resort to new expedients to earn a living, the boarding of the insane will be taken up if the way is legally opened for it. This has already proved to be the case in some parts of Massachusetts, where the experiment of boarding out on a small scale under the supervision of the Board of Lunacy and Charity has been attended, according to the opinion of the Board, with satisfactory results. An advantage derived by the public from the boarding-out of the insane is that it is relieved from the expense of providing and maintaining shelter for them. The benefits accruing to the insane are that they are thus permitted to enjoy greater freedom and the natural surroundings and associations of home life. As to the question of economy, if we take the same standard of diet and bodily comfort for both asylum and family care, with a complete and efficient method of supervision in the latter case, the cost of maintenance in families, it is believed, cannot be made much if any less than in asylums.

While it is due the insane that every reasonable requirement of humanity should be met in providing for their care and treatment, we should not overlook the rights of the sane. Therefore, in extending the boarding-out system in this country, as will undoubtedly be done, it is desirable to keep in mind the welfare of the communities among which the insane are placed and note the effects of the intermingling. A no less rigorous supervision should be maintained than in Scotland, and the selection of patients for family care should be governed by even stricter rules than those observed in that country.

POORHOUSES.

In all the countries visited, an examination was made of institutions variously designated as poorhouses, workhouses, and alms-houses. Such of them as contained insane persons, either in their ordinary wards or in separate departments, were carefully inspected, and the condition of the insane noted. It was found that the provision for their care did not reach a proper standard. The workhouse establishments in England, primarily designed for sane paupers, have limited opportunities for special treatment of the insane. In many places lunatics are not separated from other inmates, and where it is done the indoor space allotted them is contracted and the yards for recreation small. From a lack of trained attendants there is insufficient supervision, and other essentials to proper care are wanting.

Respecting the care of the insane in workhouses, the English Commissioners say: "Boards of guardians are not always ready to adopt our suggestions as to the employing of paid attendants to look after the imbecile inmates, in cases where their number appears to us to require such attention; nor can we always obtain a recognition of the need for a diet superior to that which is found sufficient for the ordinary inmates." Inspector Henley informed me that, in the fifty-eight Unions under his supervision, there was but little separate care for the insane, Birmingham being the only one having large separate wards; but that in some six of the Unions there were small wards for epileptics and troublesome cases, and that imbeciles and idiots usually occupy the common ward, where they are more likely to annoy others than to suffer themselves. Mr. Henley, who has had large experience as an inspector of workhouses, is of the opinion that workhouse provision, although in separate departments, is wholly unsuited to the needs of the insane, and not in keeping with the interests of rate-payers. He strongly favors separate asylums for this class.

If, from the examinations made, one conviction forced itself upon my mind more strongly than another, it was that poorhouse, workhouse, or alms-house care, whether in common with sane paupers or in separate departments, but under the same control and management, is not humane, and is in many ways unsatisfactory. In keeping two classes under the same management the constant tendency is to the adoption of a uniform standard of care. Provision that would be adequate to the needs of the average poorhouse inmate is quite inadequate to the necessities of the insane.

No statute can justly place common paupers and the insane on the same plane. The former come under public care, in the majority of cases, through improvidence, or physical infirmity caused by licentious or criminal ways of life; the latter reach a condition of peculiar helplessness and dependence from mental disease, frequently the result of causes for which they are not to blame. The insane are designated by Blackstone as " the victims of accidental misfortune," which cannot be said, as a rule, of the inmates of a poorhouse. It is therefore unjust that the insane should be denied special care and treatment and that they should be made to bear the stigma consequent upon classification and association with ordinary paupers.

LOCAL OR DISTRICT CARE.

In some of the States of this country there is a strong and growing public sentiment in favor of local or district care for the chronic insane. Accepting it as a probability that local care will continue to extend as population increases and the numbers of the chronic insane multiply, and having in view the fact that many populous localities have already elected to permanently provide for their chronic insane, it appears to be of very great importance that careful consideration should be given to the kind of provision made for them, and to the question of their supervision and management. I am clearly of the opinion that when large municipalities, populous counties, or congeries of counties decide to make local provision for this class, the following principles should be regarded as fundamental :

1. The chronic insane should be placed under independent, non-partisan boards of management, the members of which should be appointed by one of the courts for long terms, and should be unsalaried. These boards should be governed, as near as may be, by the same rules that govern boards of managers of State asylums.

2. Accommodation entirely separate from that for sane paupers, should be provided on considerable tracts of good land, conveniently accessible by rail or water.

3. The financial and other transactions should be kept separate from those relating to the relief of ordinary paupers.

4. The standard of care should be such as to meet the approval of State authorities.

In this way it is believed a disinterested and intelligent government of these institutions may be secured, with the advantages resulting from accumulating experience, a continuous administration, and a uniform policy. This cannot be the case under the present county system, with its frequent change of officers and the recognized custom of making institutions for the poor in a greater or less degree mediums for the distribution of political patronage.

STATE WARDS.

From the information obtained, I am led to believe that in Europe as well as in the United States the attempt of relatives in moderate circumstances to pay for the board of a member of the family in an institution for the insane at the rates charged often results in great hardship. Officials connected with public relief in some of the States of this country are aware that it frequently happens that the efforts made by a family to support one of its number in an insane asylum have completely pauperized it and brought its remaining members, including children of tender age, upon the public for support. In Illinois, Indiana, Ohio, and Minnesota, where free provision is made for the insane by the State, such cases of distress and breaking up of families do not occur. The law making the care of the insane a State charge is primarily based on the principle that when the tax-paying citizen becomes insane he is no less entitled to the special care of the State than the citizen who does not pay a tax.

In New York and some other States the asylums, while ostensibly designed for the dependent class, are open to citizens of the State who are not dependent. The maintenance and care of such become a charge against those liable for their support. This charge is in excess of that made for the care of the dependent class. Many citizens of small means are unable to bear so heavy a burden, and they should not be required to pay more than the rate charged local authorities for the maintenance and care of pauper patients. If such a rule were observed, the pecuniary distress incident to the present system would be greatly alleviated.

Respecting the policy of providing for the dependent insane, it would seem to be wise for every State to furnish free hospital treatment of the kind already described for those coming under the general term of the acute insane. If State care were free to such cases, prompt transfers to hospitals would be encouraged and the great advantages resulting from early treatment would be more generally secured.

As to the care of the dependent chronic insane in States not having a free system, it appears to me that this class might with advantage be received into State asylums from local authorities at a very liberal rate fixed by statute—one that would be less than it would cost such authorities to maintain them at a proper standard of care at home ; and that to such municipalities, counties, and districts as should elect to provide for their own quiet and harmless chronic insane a yearly per-capita allowance might be made by the State of the difference between the rate fixed by legislative enactment and the actual cost of home support, but not beyond what it had cost the State to maintain the insane in its chronic asylums, it being provided that the standard of care be verified by State authorities as equal to State asylum care, and that the insane be placed under such separate local boards as have already been recommended. It will be seen that the making of an allowance by the State in the manner proposed would be somewhat analogous to the action of Parliament in granting four shillings a week per capita to local authorities in England, Scotland, and Ireland when providing for their insane a standard of care satisfactory to the State—a form of relief that has proved highly beneficial to the dependent insane in those countries.

The extending conditionally of State aid is an acknowledgment of a continuous obligation on the part of the State, and its acceptance by local authorities is a recognition of the principle of State wardship of the insane. Deprived as they are by the laws of the State of their personal liberty, they naturally become, in one sense, her wards, and should, in their helpless and suffering condition, be so regarded. Whether they occupy buildings owned by the State or those belonging to cities or counties, the State should recognize her obligations to protect them, and should see that every requisite to their proper care is afforded them.

PUBLIC ACCOUNTS.

The reports annually presented to the English Parliament in relation to the care of the insane and other dependent classes, embody a vast amount of varied and detailed information. The keeping of accounts and the collecting and tabulating of statistics respecting relief afforded these classes, involve careful book-keeping in the charitable institutions, a considerable outlay for clerical service, and much labor on the part of public officials ; but the practical advantages derived therefrom by the Government are so great that the work is deemed indispensable.

The methods adopted by several of the American States in preparing for legislative bodies needful information relating to the care of the dependent classes, are, in some respects, superior to those of Great Britain. The system elaborated by Mr. F. H. Wines, Secretary of the Illinois State Board of Charities, is, with some slight modifications, one of the best of our American methods. In many States the lack of a general system of classifying and tabulating the expenditures of State charitable institutions is greatly felt, and the subject is one worthy the attention of legislative bodies.

It appears to me desirable that a central Bureau should classify the disbursements of all the institutions, in order to apply a uniform rule in deciding between what are ordinary and extraordinary expenditures, etc. It would seem that the Legislature of every State should require each one of its State charitable institutions to furnish a central Board or Bureau with an annual itemized statement of all its receipts and disbursements, as also an inventory of its supplies and other property in such form as the central Bureau might direct, and that this Bureau should be required to classify and tabulate the information furnished and present it yearly in printed form, at a specified date, to the Legislature. The tables should be so arranged as to be readily comparable. They should give, in addition to the financial statements usually rendered respecting the receipts and expenditures and the number of inmates in each institution, the quantity per capita of each article used, the per-capita cost of each article consumed, and the average weekly per-capita cost of maintenance, as also the average weekly per-capita cost of support of the inmates of each class of State charitable institutions.

If this system were adopted, the Legislature would be enabled to judge more intelligently of the special wants of the several State institutions and the standard of care bestowed upon their inmates. The suspicion of extravagance which occasionally, and very naturally, attaches to some institutions in States that have no general system of the kind described, would perhaps, be done away with, while others might show an excellence of management that would create a spirit of emulation tending to general improvement. By this means faults in administration would also be manifest and a remedy could be applied. Such a plan would not only prove a satisfaction to the tax-paying citizen and to the student of social science, but would greatly aid and protect boards of trustees, and officials in immediate charge.

It is not creditable to a State to be unable to show to its own citizens the interior workings of its charitable institutions and the cost in detail of maintaining the inmates. A State neglecting to adopt a comprehensive system of keeping its accounts and classifying its expenditures remains ignorant as to whether it is in advance of its neighbors or behind them in respect to the care of its dependent classes. Besides, it is losing the full benefit that might be derived from inventions and improvements that have been made in one of its institutions because they are not brought promptly and systematically to the attention of all the others.

SUPERVISION.

In nearly all the countries visited there exists some kind of supervision by the general government over the insane, either through boards of commissioners, or officers appointed for the purpose. Among the various methods that have been adopted, those of England and Scotland appear to be the best. The English system of supervision is, however, cumbersome and expensive to maintain. In the United States the care of the dependent classes is so intimately connected that some of our State Boards have been organized to supervise not only lunatics, but all classes requiring public relief. The work of these organizations has thus far been highly creditable, and they give promise of becoming more useful with increased experience.

The success of general supervising boards does not depend so much on the amount of power they possess, as upon the judicious exercise of the authority with which they are invested. This is well illustrated in the policy of the Scotch and English Lunacy Boards, through which great reforms and wholesome regulations have been brought about. Both of these bodies, though active and earnest in the advocacy of humane methods, and watchful in maintaining a strict observance of the statutes respecting the care and treatment of the insane, nevertheless have sought to carry their measures by moral suasion rather than by the arbitrary enforcement of the law. The late Chairman of the English Board said: "I am sure that the success we have had with the county asylums has been entirely because we have done every thing by persuasion, by the force of experience and constant observation."

Supervising State Boards should be so constituted as to be non-partisan, and should include members of the medical and legal professions and persons of well-known business qualifications. They should be appointed for long terms. A Board should not be so large as to be unwieldy. Its members should have definite duties to perform, including certain well-defined visitations, and it should have the power to employ competent deputies, who should be constantly active in making inspections. The Board should report annually to the Legislature, giving a full account of its work, the condition of the classes under its supervision, with suggestions and recommendations for their humane and economical care. Such Boards should be vested with the necessary statutory powers to protect fully the insane and further their interests.

In every lunacy system there should be kept by the supervising Board of Commissioners a registry of all the insane under official cognizance, and when a patient is admitted to an asylum the admission papers with the medical certificates should be forwarded to the commissioners. These should be accompanied by a statement from the asylum superintendent of the mental and physical condition of the patient. It should be practicable for commissioners to make transfers from one asylum to another, or from asylum to family care. They should also be empowered to correct errors, to make discharges after special examinations, and to prosecute for abuse of patients. In the formation of supervising Boards, the Scotch system furnishes profitable suggestions.

INDEX.

Aarhus, asylum at, 209
Accounts, public, 359
Acute Insane, provision for, 297
Admission of patients, in England, 27 ; in Scotland, 118 ; in Ireland, 180 ; in France, 199 ; in Belgium, 201 ; in Prussia, 203 ; in Province of Saxony, 281 ; in Switzerland, 220 ; (Résumé), 347
Air-space, 306
Alcoholic stimulants, 342
Alt-Scherbitz Provincial Asylum, situation, central buildings, Manor House, cottages, land, 279 ; grounds, memorial of Professor John Maurice Koeppe, object of asylum, supervision, control, management, 280 ; medical staff, officers, class of patients received, admission, 281 ; discharge, release on trial, 282 ; classification, rates charged, reception stations, observation stations, detention houses, administration building, 283 ; hospital, description of central institution buildings, 284 ; isolation rooms, 285 ; non-restraint, strong dresses, attendants, medical examination, 286 ; villas, furniture and furnishing, 287 ; hamlet, two-story pavilions, kitchen, vehicle for carrying food, 288 ; heating, water, bucket system, farm buildings, dairy, brick-yard, hot-house, fish-pond, orchard, employment, 289 ; excursions, theatrical entertainments, concerts, dances, bowling, swimming-bath, correspondence, 290 ; visits by relatives and others, a fundamental principle in treatment, 291
America, early provision for insane in, 11 ; initial point of hospital treatment in, 11 ; first State asylum in, 12 ; establishment of hospital for insane by Society of Friends in, 12
Amusements, 47, 54, 57, 63, 67, 74, 82, 89, 98, 101, 107, 183, 185, 191, 215, 223, 228, 234, 290 ; (Résumé), 329

Andrews, Dr. J. B., 324
Arthur's Seat, 168
Asylum at Jerusalem, 6
Asylums for the chronic insane, 304, 305
Attendants, 49, 51, 56, 65, 79, 93, 143, 149, 157, 163, 192, 215, 222, 223, 229, 234, 286 ; (Résumé), 322, 327

Banstead Asylum, object, Committee of Visitors, 55 ; situation, quantity of land, buildings, medical staff, officers and employees, attendants, airing-courts, 56 ; restraint, dress, tell-tale clocks, electric fire-bells, fire-brigade, general kitchen, water supply, employment, amusements, library, 57 ; dramatic entertainments, picnic excursions, asylum band, dances, chapel, cost of maintenance, 58
Barony Parochial Asylum (Lenzie), management, class of patients, Lenzie Act, situation, freedom from restraint, land, farm dwellings, 142 ; porter's lodge, buildings, grounds, officers, attendants, dormitories, 143 ; single rooms, strong rooms, infirmary wards, aviary, 144 ; bath-rooms, water distribution, sewing-rooms, entertainment hall, dances, asylum band, reading matter, dining-hall, food, 145 ; kitchen, non-restraint, 146 ; employment, tact in management, farm stock, sewerage, 147 ; laundry, dress, uniforms, 148 ; rules governing night attendants, 149 ; daily routine, 150 ; glazed corridors, chapel, Dr. Rutherford's views respecting non-restraint, 151 ; stimulating beverages, milk, 152 ; Muckroft Farm, Farm Steading, Fauldhead, cost of maintenance, 153
Bath and dressing rooms, 309
Belfast Asylum, class of patients, 184 ; chronic cases, buildings, grounds, single rooms, dormitories, dining-hall, employment, airing-courts, ball-playing, asylum band, 185

Belgium, reorganization of lunacy system in 1850, use of wooden cages forbidden, 197 ; boarding the insane in families, 200 ; Lierneux, population, number of insane, support of pauper lunatics, government and municipal asylums, private establishments, admission of patients, 201 ; visitation and inspection, discharge of patients, 202 ; Gheel (see G.).
Belt and leg-chain, 10
Bethlem Hospital, 15, 16, 18, 19
Bevan, Mrs. Elizabeth, 159
Bicêtre, 7, 195, 224
Birmingham Borough Asylums :
Winson Green, situation, buildings, grounds, management, 103 ; furnishings, padded rooms, refractory wards, restraint, seclusion, employment, uniforms, dress, 104 ; airing-courts, experience of an old superintendent, 105 ; calisthenics, singing classes, cost of maintenance, 106
Rubery Hill, object, situation, 104 ; cost of maintenance, 106
Blackstone, 17, 355
Blair, Dr. Robert, 153
Bloomingdale Asylum, 13
Boarding-Out, in Scotland, 115, 130 ; in Belgium, 200 ; in Norway, 206 ; (Résumé), 352
Board of Control (Ireland), 175, 178
Board of Guardians (England), 42
Boards of Governors (Ireland), 179
Borough Muir, 168
British specialists, criticisms by, 13
Brookwood Asylum (see Surrey County Asylum).
Brushfield, Dr., 78, 80, 81
Buildings, 297 ; objections to large mixed asylums, 298 ; need of small hospitals for the acute insane, 299 ; plan of hospital, 300 ; reception and other cottages, 301 ; connecting corridors, 302 ; provision for the chronic insane, 303 ; construction of buildings, 305
Bulckens, Dr., 250, 259
Burghölzli Cantonal Asylum, situation, medical staff, buildings, 219 ; admission, object, classification, rate charged for pauper patients, land, airing-courts, dining-halls, dormitories, single rooms, 220 ; chapel, amusement-hall, strong dresses, muffs, refractory ward, isolating cells, hot-water baths, 221 ; bucket system, attendants, heating, employment, causes of insanity in Switzerland, 222

Case, Dr., 61
Caterham Metropolitan Asylum, object, management, supervision, situation, land, buildings, 64 ; single rooms, class of patients, attendants, cost of maintenance, infirmary, 65 ; restraint and seclusion, padded rooms, dayrooms, 66 ; water supply, Turkish bath, swimming bath, kitchen, uniforms, dress of patients, asylum band, amusement-hall, theatricals, singing class, library, 67 ; outdoor sports, indoor games, walking parties, chapel, accounts, original outlay, 68
Charenton Asylum, 7, 226 ; class of patients, history of institution, buildings, chapel, the Seine, airing-courts, 226 ; medical staff, Sisters of St. Augustine, uniforms, heating, dining-rooms, wine, diet, 227 ; dormitories, hydropathic treatment, reading-room, library, billiard-room, amusement-room, refractory wards, seclusion, 228 ; restraint, rates of maintenance, attendants, single rooms, bric-a-brac, 229
Chartered or Royal Asylums (Scotland), 115
Chronic Insane, provision for, 303
Circulating Swing, 9
Clermont Asylum, 229 ; situation, distribution of patients, rates of payment, 230
Central Institution, 230 ; classification, accommodations, dormitories, infirmary, restraint, 231 ; isolating cells, restraining chairs, camisoles, airing-courts, 232 ; summer pavilion, dress, buildings, grounds, 233
Colony of Fitz-James, 233 ; billiard-room, library, furniture, attendants, dormitories, Little Chateau, laundry, 234 ; land, agriculture, 235 ; farm buildings and stockyards, farm stock, rabbits, prizes, accommodation for farm laborers, sewage system, 236 ; employment, flouring mills, forge and machine-shop, cement-mill, cider-mill, bakery, abattoir, dairy, 237
Colony of Villers, objects, open-door system, 237 ; attending church, rewards for work, 237
Clouston, Dr. T. S., 156, 313, 324, 330
Colney Hatch Asylum, 35, 45 ; situation, buildings, porter's lodge, land, class of patients, Committee of Visitors, 45 ; Victoria Fund, interior decorations, single rooms, dormitories, sitting-

rooms, 46; amusements, classification, infirmary, dining-halls, employment, 47; laundry, amusement-hall, dancing, dramatic entertainments, airing-courts, restraint, 48; seclusion, padded rooms, attendants, uniforms, heating, ventilation, chapel, 49; cost of maintenance, 50
Commission appointed by Royal College of Physicians, 17
Commission, Royal (Scotland), appointment of, 1855, investigations by, report of, 112
Commission, Royal (Ireland), 1857, appointment of, report of, 175
Commission of Inquiry (Ireland), 1879, recommendations of, 176
Commissioners, 1828, appointed to license and visit establishments in metropolitan district, 21
Commissioners in Lunacy (England), appointment of, powers of, 24; immediate jurisdiction of, granting licenses by, 26; statement of respecting unnecessary detention, 35; visitation by, 37; salaries of, expenditures of, 38; protest of against large asylums, 298; views of respecting employment, 334; respecting workhouse care, 354
Commissioners in Lunacy (Scotland), establishment of Board of, 114; powers and duties of, 117
Committee of House of Commons (1807), inquiry of, 21
Committee of House of Commons (1877), inquiry of respecting violations of personal liberty, opinions of, 33
Committee of Visitors (England), by whom appointed, 25; powers of, 25, 32
Conolly, Dr., 22. 50, 329
Conradsberg Asylum, class of patients, buildings, 206; rates charged for maintenance, classification, dress, furniture and furnishings, dormitories, refractory ward, single rooms, heating, women's work-room, industries, weaving, spinning, 207; airing-courts, chapel, dining-rooms, food, 208
Cork District Lunatic Asylum, situation, grounds, object, building, Board of Governors, medical staff, 182; house staff, single rooms, dormitories, fire-places, dining and amusement hall, airing-courts, ball-playing, 183; land, employment, restraint, padded rooms, Turkish bath, chaplains, asylum band, 184

County and Borough Asylums (England), objects of, management of, 25; cost of maintenance in, 43
Court of Chancery, 17
Coxe, Dr. James, 348
Crichton, Dr., 111
Crichton Institute, 111
Criminal Insane, 351
Criminal Lunatic Asylum (Ireland), 177
Criminal Lunatics, discharge of, 33
Cromwell, 169

Dark Ages, ignorance in the, 4
Demonology, 5
Denmark, lunacy legislation, 206; asylum at Aarhus, asylum at Vordingborg, asylum for the chronic insane at Viborg, 209; St. Hans Hospital (see S.).
Dietary and Dining-rooms, 340
Discharge, (England), 31; on trial, 36; (Scotland), 121; of dangerous lunatics, 123; on trial, on probation, 124; (Ireland), 181; (France), 199; (Belgium), 200; (Province of Saxony), 282; (Résumé), 347
District Asylums (Scotland), 115; (Ireland), 175, 176
Dix, Miss, 12, 111
Donegal District Asylum, situation, 185; gateway lodge, buildings, grounds, single rooms, associated dormitories, open fires, restraint, seclusion, airing-courts, industries, 186
Doors, 307
Dress and Clothing, 340
Dublin, asylum at (see Richmond Asylum).
Duncan, Dr. Andrew, 159

Eames, Dr., 329
Earle, Dr. Pliny, 250
Edward II., 17
Eg, asylum at, 205
Egyptians, ancient, treatment of insane by, 2
Employment, 47, 52, 57, 72, 80, 87, 92, 98, 102, 104, 147, 157, 161, 184, 185, 186, 188, 191, 206, 207, 210, 216, 218, 222, 237, 273, 289; (Résumé), 332
England, former treatment of the insane in, Bethlem Hospital, early legislation, 15; the insane wandering at large, St. Luke's Hospital, inadequate provision, public exhibition of the insane, 16; appointment of commission by Royal College of Physicians, estates of idiots, Chancery care, 17; cruel treatment of the insane,

case of Norris, 18; York Asylum, York Retreat, 19, 20; inquiry of Select Committee of the House of Commons, county and borough asylums, appointment, inquiry, and report of Metropolitan Commissioners, 21; provision for the insane prior to 1844, 22; means of restraint, Act of 1845, erection of asylums made obligatory, 23; appointment and powers of Lunacy Commissioners, building asylums made compulsory, 24; object and management of county and borough asylums, 25; maltreatment of patients, licensed houses, 26; registered hospitals, admission, 27; discharge, 31; transfer, investigation (1877) by Select Committee of House of Commons, 33; testimony of Lord Shaftesbury, 34; complaints of unnecessary detention, 35; liberation on trial, 36; visitation by Commissioners in Lunacy, 37; estimated cost of lunacy supervision, number and distribution of lunatics, 38; lunatics in workhouses, government grant, 39; private single patients, outdoor pauper lunatics, 40; metropolitan district asylums, Metropolitan Asylums Board, Local Government Board, 41; transfer of insane to workhouses, 42; supervision of workhouses, cost of maintenance in county and borough asylums, increase of pauper lunatics, 43; lunatics, paupers, population, 44 Colney Hatch Asylum (see C.); Hanwell Asylum (see H.); Banstead Asylum (see B.); Leavesden Metropolitan Asylum (see L.); Caterham Metropolitan Asylum (see C.); Sussex County Asylum (see S.); Surrey County Asylum (see S.); West Riding Asylum, Wakefield (see W.); South Yorkshire Asylum, Wadsley (see S.); Lancaster County Asylum, Prestwich, (see L.); Lancaster County Asylum, Whittingham (see L.); Birmingham Borough Asylums (see B.); Friends' Retreat (see F.)
Esquirol, 195, 226, 228, 241, 277

Fire, protection against, 311
Fire-places, 307, 310
Fitz-James, colony of (see Clermont).
Flodden Field, 168
Foundation Walls, 306
Foville, Achille, 200
France, cruel treatment of the insane in the latter part of the 18th and early part of the 19th centuries, 195;
present provision for the insane, 197; establishment of asylums made obligatory, inspection of public and private asylums, supervision of public asylums, 198; official visitation of asylums, admission, discharge, 199; right of reclamation, opinion of Achille Foville, 200; La Salpêtrière, (see S.); Asylum of Ste. Anne (see S.); Asylum at Charenton (see C.); Clermont en Oise (see C.).
Francis, Dr. Lloyd, 335
Frankford, hospital at, 12
Fraser, Dr., 139
Freedom, greater, 317
Friedrichsberg Asylum, government, porter's lodge, Pensionnat, principal building, park and cultivated grounds, class of patients, 212; variety of accommodation, routine of meals, beer, smoking-room, bath-rooms, 213; library, dining-rooms, dormitories, 214; clothing, valuables, letters, heating, lighting, bowling-green, billiard-room, music hall, chapel, attendants, salaries, 215; solitary rooms, refractory wards, asphalt floors, chronic insane, two-story buildings, yards, non-restraint, employment, farm stock, 216; farm buildings, classification, 217
Friends' Retreat, 19, 20, 22, 106; non-restraint, William Tuke, Society of Friends, 106; situation, grounds, management, income, buildings, The Lodge, Bellevue House, villa residence, interior, sitting-room, religious services, association rooms, lectures and entertainments, 107; rates of payment, cost of support, padded rooms, strong garments, seaside resort, 108
Friends' Retreat, Penn., 351
Furnishing and Decoration, 312

Gaustad Asylum, 204; situation, buildings, heating, dormitories, airing-courts, rooms for the refractory, douche, 205; padded rooms, strong dresses, employment, 206
George II., 15
George III., attacked by insanity, 16
Germany, "mad-houses" in (1803), 194; curative treatment in, 197; asylum care, family care, public and private asylums, open asylums (*Offene Anstalten*), 203; Friedrichsberg Asylum (see F.); Provincial Asylum, Halle (see H.); Provincial Asylum, Alt-Scherbitz (see A.).

INDEX.

Gheel, 3, 204; town of, 239; Church of St. Dymphna, 240; religious ceremonies, 241; legend of St. Dymphna, 242; commune of Gheel, 243; general description of, 244; the Infirmary, 245; supervision and medical visitation of the insane, 246; inspectors, description of Infirmary, 247; means of restraint, Hôtes and Nourriciers, 249; treatment of the insane prior to 1851, 250; class of patients, rates of payment, 251; inspection of town dwellings, 253; inspection of cottages in the country, 263; regulations regarding labor, etc., rewards for work, 273; advantages and disadvantages, 274

Gifts of land, 296
Glen-na-galt, 3
Gloucester, 22
Government Grant (England) 39; (Scotland), 127; (Ireland), 177; 358
Greek Medical School, 2
Grounds, 314

Halle Provincial Asylum, class of patients, government, location, buildings, land, employment, improvements, sewerage, gas and water pipes, 218; farm cottage, infirmary, sanitary arrangements, unlocked doors, associated dormitories, single rooms, 219

Hamburg, asylum near (see Friedrichsberg)

Hanwell Asylum, class of patients, situation, buildings, land, 50; hospital for infectious diseases, attendants, uniforms, ventilation, thermometers, proportion of single rooms, decorations, furniture, 51; employment, 52; restraint, strong dresses, padded rooms, 53; seclusion, airing-courts, clothing, amusement-hall, dancing parties, outdoor games, picnics, reading matter, religious services, cost of maintenance, 54

Haywards Heath, asylum at (see Sussex County Asylum)
Henley, Inspector, 42, 354
Henry VIII., 15
Hill, Dr. R. Gardiner, 18, 22, 110
Holy Wells, 3
Hospice St. Vincent de Paul, 178
Hospitals, 299, 304
Hôtel Dieu, 196
Houses of Industry (Ireland), 174

Idiot, definition of term, 17

Infirmary Accommodation, 307
Insane, The, need of public provision for, 1; treatment of by ancient Egyptians, supposed to be possessed by demons, 2; provision for by Mohammedans, provision for and treatment of in Spain, 6; exhibited in cages, 6, 16; cruel treatment of, 2, 6, 18, 110, 173, 194, 195

Insanity, in the Middle Ages, 2; strange devices for treatment of, 8
Inspectors (Ireland), 178, 179
Ireland, Dean Swift, St. Patrick's Hospital, neglect of the insane, 172; cruel treatment, lack of asylum accommodation, 173; governmental inquiry, 1810, asylum built in Dublin, Houses of Industry, successive Parliamentary inquiries, Act of 1821, legal status of the insane prior to 1821, provisions of the Act of 1821, formation of districts, 174; establishment of Board of Control, asylums built, government loan, expenditures, appointment of Royal Commission 1857, report of same, 175; recommendations of Commission of Inquiry 1879, lunatics at large, population, number and distribution of lunatics, 176; cost of maintenance in district asylums, government grant, cost of maintenance in poorhouses, inadequate provision for the chronic insane, government of lunatics in poorhouses, Local Government Board, 177; condition of poorhouses, private asylums and licensed establishments, Board of Control, Inspectors, 178; visitation by Inspectors and reports of, Boards of Governors, appointment of medical officers, granting licenses, 179; dangerous lunatics, admission, 180; discharge, insufficiency of land, 181; industries, 182; Cork District Asylum (see C.); Belfast Asylum (see B.); Donegal District Asylum (see D.); Richmond District Asylum (see R.)

James IV., 168
James V., 169
James VI. of Scotland, 5
Jerusalem, asylum at, 6
Justices (England), required to provide asylums, 23, 24; required to appoint committees to build asylums and committees to manage same, 25; empowered to grant licenses, 26; required to appoint Visitors to licensed houses, 27; order of, 28

Labitte, Dr. George, 230
Labitte, Dr. Gustave, 230, 236
Lalor, Dr. Joseph, 106, 187, 188, 193
Lancaster County, population, paupers, pauper lunatics, 94
Lancaster County Asylum (Prestwich), situation, buildings, 94; porter's lodge, dress of patients, annexe, infirmary wards, convalescent wards, 95; ward for refractory female patients, single rooms, heating, dormitory, padded rooms, seclusion, 96; dining-halls, entertainment hall, use of beer, 97; land, employment, farmhouse, detached cottage, walking parties, dancing, theatrical entertainments, chapel, causes of insanity of inmates, 98
Lancaster County Asylum (Whittingham), situation, buildings, land, annexe, dormitories, thermometers, 100; day-rooms, padded rooms, non-restraint, strong dresses, wet packing, dining and amusement hall, airing-courts, walking parties, outdoor recreation, theatricals, 101; concerts, employment, washing by hand, cost of maintenance, heating, 102; religious services, 103
Land for institutions for the insane, 294
Lawson, Dr., 140
Leavesden Metropolitan Asylum, situation, outlay per capita, management, 58; officers and employees, buildings, conservatories, 59; heating, description of interior, electric clocks, 60; hydropathic treatment, 61; laundry, water supply, sewage, restraint, padded rooms, seclusion, recreation hall, 62; amusements, costume ball, walking parties, dress of patients, chapel, library, 63; cost of maintenance, 64
Lenzie, asylum at (see Barony Parochial Asylum)
Letterkenny, asylum at (see Donegal Asylum)
Letters, 290, 344
Ley, Dr., 95, 98, 326
Licensed Houses (England), 26; Visitors of, medical attendance at, 27; order for admission to, notice of admission to, 29, 30; (Scotland), condition of insane in, 113; number of, 115; medical visitation of, 126; (Ireland), 175, 179; (France), establishment of, 198; (Belgium), establishment of, 201
Lierneux, insane colony at, 201
Lincoln Asylum, 22

Local Government Board (England), supervision by, salaries of Inspectors of, 38; central authority in poor-law matters, reports submitted to, 43
Local Government Board (Ireland), 177
Local or District Care, 356
Location, 293
Lochmanur, 3
Loch Maree, well of, 4
Lord Chancellor, appointments by, 38
Lunacy Legislation (England), 1744, 15; 1774, 17; 1808, 1828, 1832, 1842, 21; Act of 1845, 23, 34, 36; Act of 1853, 24, 30, 31; Act of 1862, 30, 37, 42; 1868, 42; (Scotland), 1815, 1829, 1841, 111; Act of 1857, 114; 1862, 1866, 120, 124; 1880, 120; (Ireland), 1821, 174; (Belgium), 1850, 197; (Norway), (Sweden), (Denmark), 206
Lunatic, definition of term, 17, 24
Lunatics, estates of, 17

Macdonald, Dr. A. E., 309
MacDonald, Dr. Carlos F., 352
McLean Asylum, 12, 351
Mary Queen of Scots, 168
Masters in Lunacy, salaries of, 38
Medical Certificates, 27, 28, 29, 30, 31, 37, 110, 120, 122, 123, 129, 179, 180, 199, 201, 203, 281
Medical Officers, 321
Menstone, asylum at, 83, 302
Metropolitan Asylums Board, 41, 58
Metropolitan Commissioners in Lunacy, appointment of, report of, 21
Metropolitan District Asylums, 41
Middle Ages, insanity in, 2; practice of medicine by monks in, 4; asylums in, 6
Middlesex County, area, population, paupers, 45
Mid-Lothian and Peebles District Asylum, situation, management, land, buildings, grounds, cottages, private patients, rates for maintenance, dormitories, 154; single rooms, heating, ventilation, lighting, day-rooms, 155; billiards, lavatory, infirmary wards, halls and corridors, dining-hall, food, chapel, 156; padded rooms, non-restraint, attendants, employment, dress, 157
Mitchell, Dr., 93, 339
Mitchell, Dr., Commissioner, 121
Mohammedans, provision for insane by, 6
Monks at Saragossa, 6
Morningside, asylum at (see Royal Edinburgh Asylum)
Murray's Royal Asylum, 111

INDEX. 371

New York, State provision for insane in, 12
New York Hospital, 12
Night Attendants, rules governing, 149
Night Service, 327
Non-Restraint, 22, 50, 78, 89, 101, 105, 106, 146, 151, 157, 216, 286; (Résumé), 319
Norman, Dr. Conolly, 193
Norris, cruel treatment of, 18
Norway, government of State asylums, government of local asylums, boarding-out, State wards, lunacy legislation, 206; Gaustad Asylum (see G.)
Nursery for Plants, 80, 315

Orderly Arrangement, 315
Outdoor Pauper Lunatics, 40

Padded Rooms, 49, 53, 62, 66, 71, 86, 91, 96, 101, 104, 108, 157, 163, 184, 188, 206, 225; (Résumé), 319
Paetz, Dr., 281, 286, 290
Parigot, Dr., 250
Paris, 7; early treatment of insane in, 195; asylums at, 222, 224, 226
Pariset, 195
Parliamentary Investigations, 21
Parochial Asylums (Scotland), 115
Patients, law respecting maltreatment of, 26
Pauper, definition of term, 44
Pauper Lunatics, increase of, 43
Peeters, Dr., 240, 246
Pennsylvania, first hospital for insane in, 11
Pensions, 71, 84, 325
Perceval, Secretary, statements of respecting admission to asylums, 28; undue detention, 32; outdoor pauper lunatics, 40
Petit, Dr. Joseph, 186, 318
Pinel, 7, 19, 22, 195, 196, 224
Pipes, supply and waste, 308
Poorhouses, (Scotland), 115, 127; (Ireland), 177; condition of, 178; (Résumé), 354
Post-mortem Examinations, 90, 93; (Résumé), 346
Prestwich, asylum at (see Lancaster County Asylum)
Private Asylums (Scotland), 115; (Ireland), 178
Private Dwellings (Scotland), insane in, 115
Private Single Patients, 40
Prussia, admission to asylums in, private asylums in, 203; maintenance of the incurable pauper insane, dangerous cases, 204; Provincial Asylum (see Halle); Provincial Asylum (see Alt-Scherbitz)
Psychology, study by ancient Greeks of, 4
Public Accounts, 359

Rayner, Dr., 51, 53, 298
Reforms in continental countries, 196
Registered hospitals, 27; medical attendance at, 27
Religious Exercises, 328
Remuneration, 72, 81, 237, 273; (Résumé), 338
Restraint, 23, 48, 53, 57, 62, 66, 72, 163, 184, 186, 225, 229, 231, 249
Résumé, 293; Location, 293; Buildings, 297; Furnishing and Decoration, 312; Grounds, 314; Orderly Arrangement, 315; Sewage, 316; Greater Freedom, 317; Non-Restraint, 319; Medical Officers, 321; Attendants, 322; Night Service, 327; Religious Exercises, 328; Amusements, 329; Employment, 332; Remuneration, 338; Dress and Clothing, 340; Dietary and Dining-rooms, 340; Alcoholic Stimulants, 342; Social Distinctions, 342; Visitation, 343; Letters, 344; Post-mortem Examinations, 346; Admission and Discharge, 347; Voluntary Patients, 350; Summer Resorts, 351; Criminal Insane, 351; Boarding-out, 352; Poorhouses, 354; Local or District Care, 356; State Wards, 357; Public Accounts, 359; Supervision, 361
Retreat, The, 178
Rheinau, asylum at, 220
Richmond District Asylum, medical staff, attendants, grounds, building, dining-hall, food, 187; associated dormitories, padded rooms, day-rooms, employment, school system, 188; laundry, 190; table showing employments, amusements, walking parties, 191; seclusion, dress, attendants, salaries, 192; churches, abolishment of airing-courts, 193.
Riel, 194
Rœskilde Fjord, 209
Ross, Captain David, 109
Rotvold, asylum at, 205
Royal College of Physicians, 17
Royal Edinburgh Asylum, situation, management, rates for maintenance, charity fund, 158; West House, grounds, cricket club, food, ale, dining-halls, 159; general kitchen, diet

kitchens, recreation hall, heating, 160; shoe-house, employment, printing-office, 161; *Morningside Mirror*, laundry, dormitories, single rooms, open-door system, airing-courts abolished, 162; restraint, seclusion, padded rooms, trained attendants, 163; seaside resort, East House, furniture and furnishings, open fires, billiards, reading-rooms, books, 164; additions and improvements, 165; cottage for female patients, Myreside Cottage, 167; Craig House, sycamores, extended views, seat of the Carmichaels, 168; furniture and furnishings, dining-room, 169; drawing-rooms, ornamental grounds, class of patients received, freedom from restraint, convalescing patients, 170
Rush, Dr., 11
Rutherford, Dr., 146, 151, 152, 153

St. Andrew's Hospital, 335
St. Anne Asylum, situation, buildings, object, medical staff, 224; Sisters of St. Joseph, day servants, refractory wards, isolating cells, restraining chairs, padded rooms, airing-courts, roofed galleries, heating, 225; hydropathic treatment, 226
St. Dymphna, 3; legend of, 242
St. Fillan, wells of, 4
St. Hans Hospital, near Roeskilde, 208; class of patients, supervision, classification, buildings, grounds, Roeskilde Fjord, single rooms, associated dormitories, 209; seclusion, dining-halls, kitchens, food wagons, chapel, billiard-room, bowling-alley, reading matter, employment, knitting, 210; asylum waste, beach bathing, 211
St. Luke's Hospital, 16, 346
St. Marce, well of, 4
St. Nun's Pool, 3
St. Patrick's Hospital, 172, 178
St. Winifred, well of, 3
Salpêtrière, 8, 196, 222; object, buildings, class of patients, school, 222; attendants, cottages, dormitories, cells, douche, entertainments, kitchen, 223; wine, grapes, dining-hall, Pinel, Bicêtre, 224
Sanborn, F. B., 346
Saragossa, monks at, 6
Schools, 74, 106, 188, 223
Scotland, establishment of present Lunacy System, superstitions, witchcraft, sentenced to death for witchcraft, 109; imprisonment at Inverness, inadequate provision, private aid, Royal Edinburgh Asylum, 110; Murray's Royal Asylum, Crichton Institute, chartered or royal asylums, government aid, lunacy legislation in 1815, 1829, 1841, p. 111; appointment of Royal Commission in 1855, investigation by same, report of, 112; recommendations of Commission, important legislation of 1857, creation of Board of Lunacy, present provision for the insane, 114; royal asylums, district asylums, parochial asylums, lunatic wards of poorhouses, private asylums, insane in private dwellings, 115; desirable reforms, General Board of Commissioners in Lunacy, 116; Deputy Commissioners, powers and duties of the Lunacy Board, 117; admission of patients, 118; care of property of the insane, voluntary patients, 120; discharge of patients, 121; limitation of sheriff's order, 122; dangerous lunatics, 123; liberation on trial, probationary discharge, 124; penalty for maltreatment of patients, medical attendance in licensed establishments, 125; proportion of insane to population, increase of the insane, distribution of the insane, how maintained, cost of maintenance, 126; government grant, poorhouse care, exposition of Lunacy System, 127; boarding-out, 130; correspondence, 132; rates for boarding-out, 133; inspection of insane in private dwellings, 134; Barony Parochial Asylum (see B.); Mid-Lothian and Peebles District Asylum (see M.); Royal Edinburgh Asylum (see R.)
Scott, Hon. Francis, 298
Scott, Sir Walter, 4, 109, 168
Seclusion, 49, 62, 66, 72, 78, 89, 96, 104, 163, 192, 210, 216, 221, 225, 228, 232
Sewage, 316
Sewers, 308
Shaftesbury, Earl of, testimony of relating to abuses in asylums, 22; statement of respecting discharge, 34; 298, 321, 344, 362
Shakespeare, 16
Sibbald, Dr., 127, 131
Sites for institutions for the insane, 293
Smith, Dr. Stephen, 324
Social Distinctions, 342
Society of Friends, 12, 19, 20, 106, 178

South Yorkshire Asylum, situation, buildings, porter's lodge, church, greenhouse, day-rooms, 90 ; men's infirmary, refractory wards, strong rooms, padded rooms, heating, 91 ; bathing arrangements, food, kitchen, bakery, brewery, beer, laundry, water supply, employment, 92 ; land, farm cottages, hospital accommodation, post-mortem examinations, attendants, uniforms, airing-courts, 93 ; dining and recreation hall, library, chronic cases, cost of maintenance, 94
Spain, provision for insane in, treatment of insane in, 6
Spiritual Agencies, 3
State Aid, 39, 127, 177, 357, 358
State Provision for the Insane, 12
State Wards, 357
Steenberg, Dr., 211
Stewart Institution, 178
Summer Resorts, 351
Superstitions, 4, 109
Supervision, 301
Surrey County, population, paupers, pauper lunatics, 75
Surrey County Asylum (Brookwood), management, situation, buildings, 75 ; interior decorations, dining-rooms, heating and ventilation, kitchens, room for new arrivals, day-rooms dormitories, 76 ; temperature, single rooms, bath-room, 77 ; cottage hospital, religious services, cost of maintenance, non-restraint, seclusion, 78 ; padded rooms, classification, uniforms, attendants, salaries, 79 ; land, employment, nursery, conservatory, 80 ; washing by hand, remuneration, 81 ; recreation and amusement, recreation hall, fancy-dress balls, musical meetings, printing, library, 82
Sussex County, population, paupers, pauper lunatics, 69
Sussex County Asylum (Haywards Heath), opening, management, situation, ornamental grounds, quantity of land, class of patients, buildings, 69 ; dormitories, wards for better class female patients, infirmary ward, suicidal and epileptic dormitories, 70 ; padded rooms, detached hospital, dress of patients, uniforms, wages, pension, kitchen, dining-room for female patients, 71 ; beer, dietary, rewards for work, water supply, fire-hose, sewerage, wet and dry packing, locked dresses, seclusion, employment, 72 ; skilled artisans, officers and employees, cost of maintenance, reading matter, 73 , school, religious services, walking parties, drives, picnic excursions, outdoor games, amusement hall, theatricals, concerts, fancy-dress ball, rules, 74
Sweden, lunacy legislation, 206 ; control of asylums, 208 ; Conradsberg Asylum (see C.).
Swift, Dean, 172, 174, 178
Swimming Baths, 67, 211, 290, 309
Switzerland, Burghölzli Cantonal Asylum (See B.)

Thermometers, 311
Training Schools, 324
Transfers, 33
Tuke, Dr. Batty, 331
Tuke, Dr. D. Hack, 342
Tuke, William, 19, 22, 106

United States, progress of reform in, 13

Valley of Lunatics, 3
Ventilation, 310
Viborg, asylum at, 209
Vienna, early asylum at, 197
Villers (see Clermont).
Virginia, first asylum in, 12
Visitation, 343
Visitors of Chancery patients, 38
Voluntary Patients (Scotland), 120 ; (Résumé), 350
Vordingborg, asylum at, 209

Wadsley (see South Yorkshire Asylum)
Wakefield (see West Riding Asylum)
Wallis, Dr., 102, 324
West Riding, population, paupers, pauper lunatics, 83
West Riding Asylum, buildings, land, medical officers and attendants, 83 ; salaries, uniforms, pension, day-room, dress, dormitory, 84 ; strong dresses, locked boots, refractory ward, bath-room, heating, building for convalescing patients, 85 ; open fires, day-rooms, conservatory, single rooms, strong rooms, padded room, fire-brigade, 86 ; dining-room and recreation hall, food, general kitchen, employment, weaving, tailoring, 87 ; book-bindery, laundry, 88 ; gas, water supply, sewerage, recreation, religious exercises, entertainment, seclusion, non-restraint, 89 ; post-mortem examinations, cost of maintenance, 90

White House Asylum (Bethnal Green), abuses in, 23
Whittier, 4
Whittingham (see Lancaster County Asylum)
Willard Asylum, 295
Williams, Dr., 333
Williamsburg, asylum at, 12
Windows, 307
Wines, F. H., 359
Witchcraft, 5, 109

Woodilee (see Barony Parochial Asylum)
Workhouses, lunatics in, 42, 354
Workshops, 308

York, Friends' Retreat at, 106
York Asylum, 19; abuses in, 20
Yorkshire, population, paupers, pauper lunatics, 83

Zurich, asylum near, 219

www.ingramcontent.com/pod-product-compliance
Lightning Source LLC
Chambersburg PA
CBHW030542300426
44111CB00009B/828